it's
IN the BAG

it's IN the BAG

A New, Easy, Affordable, and Doable Approach to Food Storage

MICHELLE & TRENT SNOW

CFI
Springville, Utah

ISBN 13: 978-1-59955-385-6

Published by CFI, an imprint of Cedar Fort, Inc., 2373 W. 700 S., Springville, UT 84663
Distributed by Cedar Fort, Inc. www.cedarfort.com

LIBRARY OF CONGRESS CATALOGING-IN-PUBLICATION DATA

Snow, Michelle, 1961-
  It's in the bag : a new approach to food storage / Michelle and Trent Snow;
food photography by Stacey Pointer.
    p. cm.
  Includes index.
  ISBN 978-1-59955-385-6
  1. Food--Storage. I. Snow, Trent. II. Pointer, Stacey. III. Title.

TX601.S66 2010
641.4--dc22

2010004643

Cover design by Tanya Quinlan
Cover design © 2010 by Lyle Mortimer
Edited and typeset by Megan E. Welton

Food photography by Stacey Pointer

Printed in the United States of America

10 9 8 7 6 5 4 3 2

Printed on acid-free paper

Acknowledgments

We would like to thank our daughter, Rachel Snow, and the skilled employees at Cedar Fort for their assistance in publishing this book.

Contents

SECTION THREE: THE NITTY GRITTY

Preface

Although only two names appear on the cover of *It's in the Bag*, this book would not have been possible without my "mothers," for they taught me to live a self-reliant life style through their example. I have been blessed with three mothers, Dianne Veinotte, my mother; Grandma Eunice Garbett, my paternal grandmother; and Mary Alvey Swain, my maternal grandmother—otherwise known as Gram.

Dianne Veinotte, my mother, taught me the importance of making food storage a priority. I remember her carefully calculating how many cans of dehydrated foods our family would need for a two-year supply. I remember filling empty soda bottles with water and lining our garage walls with them. I remember filling buckets with bulk wheat. I remember Mom telling Dad, Mick Garbett, that the freezer was low on fish and game. Come hunting season, my dad would refill the freezer with salmon, pheasant, and venison. The house I grew up in didn't have an office; because of this, my parents would budget my dad's paycheck on the kitchen table as my brother and I would play board games. I remember my mother, with a calculator and the checkbook in hand, furiously working away on the budget and then announcing to my dad that they had such and such amount for food storage this paycheck. My mom's example and my dad's support taught me that food storage is a family matter—one that requires both partners to work together to provide family food security and to set the example for the next generation.

Grandma Garbett taught me how to make jam, bread, homemade ice cream, and fruit cake so delicious that you eat it rather than recycle it at the Christmas gift exchange. Grandma Garbett seemed to always have a jar of raspberry freezer jam, butter, and a hot loaf of bread or a fresh rhubarb pie cooling when I went to visit. I have wonderful memories of Grandma telling me pioneer stories while eating delicious food. The time would pass all too quickly as I ate and she talked. Grandma Garbett taught me that good food paves the way for great conversations and happy memories.

My Gram always had a huge garden, a beef, a hog, and chickens. She would harvest fruit and vegetables and home-can them from late summer through early fall. She made most things from scratch —including baked goods. My Gram left me the legacy of cooking, sewing, bottling fruits and vegetables, and making main dishes and baked goods from scratch. I would call her from 1,200 miles away and say, "Gram, how many tablespoons are in a half cup?" Or, "Gram, my bread could be used for a door stop; what did I do wrong?" Or "Gram, the recipe calls for baking powder and I am out. Is there anything I can substitute?" Gram would always know exactly what I needed to do, substitute, or how to correct the problem next time. If you were to go down into my Gram's basement, you would have thought you were in a grocery store. You would have seen row after row of beautifully canned fruits, vegetables, meats, barrels of wheat, and water as well as shelves full of toiletries and paper goods. Gram truly left me a legacy.

I have been greatly blessed to have had three mothers who not only believed in self-reliant living, but who also demonstrated that self-reliant living is daily living.

Introduction

In 2007, the idea of creating a self-reliant living group came to mind. My friend, Laura Lee Andersen, liked the idea and agreed to co-chair the group. Our self-reliant living group was open to all and met once a month. Each month, the instructor would lecture on a preselected topic, for instance, how to home can meat. The instructor would also be responsible for developing handouts and creating a tasting table. "Tasting table" was our term for food samples of all the recipes demonstrated during the presentation. We felt that if the women and men of our group could taste the recipes, they would be more likely to include them in their regular menu planning and food storage.

The idea caught on, and soon our little group of two grew to over thirty participants. Laura Lee and I set a twelve-month calendar of topics and would find instructors who were subject matter experts for each monthly topic. One month, my friend Belinda Craft wondered if our group would like to learn about complete dinners in a bag. Belinda had attended a class called "Dinner: It's in the Bag," taught by Christine Van Wagenen. Christine is the owner of the Wooden Spoon Cooking School and Family Life Center located in American Fork, Utah. Belinda taught our group the dinner in a bag idea and gave us seven recipes to get our group started. I liked the idea so much that I developed over a hundred recipes and customized the dinner in a bag concept into a self-reliant living system. That is where this book begins.

What You Will Learn from This Book

In section one of *It's in the Bag*, you will:

1. Learn how to improve and maintain your family's food storage with the convenient, economical, and spacing-saving food storage system of bag meals.
2. Have the convenience of having over a hundred proven bag meal breakfast and dinner recipes at your fingertips.

In section two of *It's in the Bag*, you will learn skills that will enable you and your family to have a more self-reliant lifestyle by using economical and nutritious food production and preservation techniques. For example you will learn:

1. How to home can fruits, vegetables, beans, beef, poultry, and fish used in bag meal recipes.
2. How to sprout seeds and beans.
3. How to convert dehydrated ingredients in a bag meal recipe into fresh food equivalents. For example, one tablespoon of dehydrated celery equals one stalk of fresh celery.
4. How to build Trent's simple and economical sprout drainage stand and suburban chicken coop.
5. How to make a yogurt that, in my opinion, has a superior taste and texture to that of commercially manufactured yogurt, for about the cost of a gallon of milk.
6. How to make butter, buttermilk, sour cream, and a variety of soft cheeses.
7. But most important, you will learn that food storage and self-reliance can be part of everyday living!

SECTION ONE:
BAG MEALS

My Bag Meal System

My bag meal system is simple. Everything you need to prepare a complete meal is in the bag. Everything means everything; for example, meat, vegetables, spices, fruit, water to cook pasta or rice, and the recipe are all included in the bag (see photographs on page 14). There are however, a few recipes that require pantry items such as peanut butter, vinegar, or vegetable oil. I place all of the dry or loose items such as rice or spices in resealable sandwich or snack bags. A sandwich bag holds a little over two cups. A snack bag holds one cup. The ingredients then go in the bag.

You may be saying to yourself, "That bag must be huge!" The bag that I use is a square-bottomed 8" x 5" x 10" plastic gift bag with ¾" wide handles for carrying comfort. On the outside of the bag, the recipe is slipped inside of a plastic CD sleeve or one with a clear window. Next to the CD sleeve is a ¾" removable color-coded label. On the label I write the expiration date, which really is the expiration date of the first item to expire in that particular bag meal (see photographs on page 15). The reusable bags cost about eighteen cents each, and the CD sleeves range from one to eighteen cents, depending on the materials and if they have adhesive backing.

Your next thought is more than likely, "That's a small bag. How many servings will a typical bag meal make?" The bag meal recipes make six to eight servings.

Twelve advantages of Bag Meals

1. You'll have organized, complete meals you know your family will enjoy.

Remember, with my bag meal system, you are storing what you eat and eating what you store. That means no waste of money, time, space, or resources, and best of all, no complaining from your family that they don't like the food. In each dinner bag meal, I not only include all of the ingredients for the entrée, but also, cooking water, a vegetable, and a can of fruit. The fruit is our dessert. Our family likes home-canned peaches, pears, applesauce, apricots, sweetened rhubarb, and plums. If you are not a home canner or dehydrator, you can purchase a variety of canned or dehydrated beef, poultry, fish, fruits, and vegetables at your local market. In the winter, we sprout an assortment of seeds and beans for winter salad fixings, and in the summer we have a salad garden growing in planter boxes along with other nutritious fruits and vegetables.

To add variety to the breakfast bag meals, I include an occasional pint of home-canned sausage, quart of fruit juice, brown sugar for hot cereal, powdered milk, and sometimes a plastic resealable sandwich bag full of home or store bought nuts and dried fruits, such as almonds, pumpkin seeds, pine nuts, sesame seeds, sunflower seeds, cherries, peaches, pears, bananas, apricots, or sweetened cranberries.

2. Bag Meals save you time and money.

Sitting down weekly to write out a menu is time consuming. Planning your weekly menu, checking the pantry, and writing your grocery shopping list adds time and frustration to your daily schedule. With my bag meal system, you can choose within seconds the meal you would like to eat that very day! I literally only spend fifteen minutes each week making my shopping list from my empty bag meal recipe cards.

Another plus of bag meals is that I only buy what I need. There is no need to second guess what I need and what my family will want to eat five days down the road. Why? Because five days down the road, they simply walk to the pantry and take the meal they want to eat off the shelf.

Bag meals have substantially decreased our grocery budget. You might be surprised to learn that before implementing my bag meal system, I was spending $520 per month for groceries. After implementing my bag meal system, my 2009 grocery budget was $250 per month, with $100 of my monthly food budget going into a savings account to purchase a locally raised, grass fed, antibiotic and hormone free beef.

To stretch my food dollars even further, my husband made a spreadsheet consisting of the bag meal recipes I want for a two-year supply, the number of times I want to eat a given bag meal in a year's time, and all of the ingredients for the bag meal recipes. This spreadsheet allows us to calculate the bulk quantity of every ingredient that we will use in our bag meals, so when there is a case-lot sale at the grocery store, I know exactly how many I need to refill my two year supply of bag meals.

3. Bag Meals have shorter prep time.

Most bag meals are ready in twenty minutes or less, start to finish. The bag meal recipes were designed to be quick. We all have been taught that during times of hardship, heating and cooking energy is expected to be scarce and at a premium price; therefore, most bag meal recipes only require foods to be combined and heated to enhance your family's mealtime pleasure; but remember, if there is no method for heating the bag meal, many bag meals can be safely eaten cold, because they contain no raw meats or eggs. The exceptions, of course, are entrees that include pasta or rice, as these must be cooked in boiling water. Home canning of beef, poultry, and fish is safe, easy, and delicious. Home-canned meats provide our family with an extra sense of security just in case the freezer should lose electricity. You will find recipes for home canning roast beef, hamburger, sausage, poultry, and fish in Section 2: Cool Stuff Your Mama Never Taught You.

4. Bag Meals will save you space.

Before utilizing my bag meal system, I had food storage ingredients organized by category in our unfinished basement. It took about a thousand square feet for all the boxes and barrels. Now, because of my bag meal system, I have 422 complete meals in a room 16 ½ feet long by four feet wide (see photograph on page 16). And believe it or not, half of the four feet width is walking space. If you do not have storage space, you might decide to store 86 bag meals under your queen-size bed. Multiply 86 bag meals by three beds, and you have 258 meals that will feed a family of six. Because of space saving bag meals, I now have room in my basement for that mother-in-law apartment I have always wanted.

5. Bag Meals offer a sense of security when you know that you'll always have all the ingredients for your meals.

Let's say that I have 365 dinner and 365 breakfast bag meals. In hard economic times or during a disaster, I know my family will have at least two complete meals a day.

The healthy adults in our family have experimented with our eating times. I use the word "adult," because children need to eat more frequently than adults. Their small stomachs are not able to hold all of the food that is required for their growing and active bodies. I write "healthy" adults because we have an extended family member who has hypoglycemia, and as such, she requires small, frequent, protein based meals. Due to her special dietary needs, our family has included a variety of extra foods that she can eat for snacks when she comes to visit.

Here's the process we went through in developing our "crisis" two-meals-a-day schedule. First, we tried breakfast at 9 a.m. and dinner at 4 p.m., but this didn't work for us. We found if we ate breakfast at 9 a.m. we were hungry around 2 p.m., so, we had dinner at 2 p.m. Predictably, we then found ourselves hungry for a second dinner at 8 p.m. Finally what worked for us was to eat our breakfast at 8 a.m., a large meal or dinner at 1 p.m. and a light snack around 7 p.m. At last we were satisfied. Our snack would consist of leftovers from dinner, bread and soup, or fruit, granola, and milk, and so on. Because each family has unique needs, I would strongly encourage families to plan their meals and "crisis" mealtime schedule according to their family members' ages and health needs before there is a scarcity of food and resources.

6. Anyone in your family who has basic reading skills can prepare a Bag Meal—even a husband who doesn't cook.

One summer, when the bag meal idea was in its infancy, I went on a five-day trip with an old friend from high school. As I was packing, my husband, Trent, asked, "Honey, what am I going to eat while you are gone?" My cheery reply was, "Sweetie, fix a bag meal." I was secretly excited, because finally, I had a bag meal guinea pig. I was going to be able to put my bag meal system to the ultimate test! When I returned home, Trent told me that the bag meals were fast, easy to make, and delicious. He was so sold on the bag meal system that he gave a bag meal testimonial at our next self-reliant living meeting.

7. All ingredients found in Bag Meals are pre-measured.

No more worries about the kids spilling as they measure ingredients. And best of all, you have all of the ingredients, including cooking water and spices, for your entire meal (see photograph on page 14).

8. The Bag Meal System is an easy way to "baby step" your way to building food storage.

I started my bag meal system by selecting a bag meal recipe

and calculating how many bags I could afford to make while stay-ing within my weekly grocery budget. Not only did I start small by adding a complete bag meal or two each week to my grocery budget, but I also took any unexpected money or salary raises and set it aside for bag meal ingredients. My only rule was that I would buy the ingredients for one complete bag meal each week. As the weeks passed, I would tell Trent how many days we could eat like a king during hard times. One day, I proudly announced that we would eat like royalty for a full year. In a matter of a few months, I had a year supply of breakfast and dinner bag meals. It is sur-prising how buying a few bag meals each week really adds to your food storage and doesn't seem to affect your grocery budget.

9. It's easy to rotate your food storage, and, best of all, you know exactly what to buy to replace your food storage ingredients.

Our family's menu planning is simple. Each night, a differ-ent person enjoys the privilege of choosing a bag meal for dinner. The empty bag, including baggies and water bottles, is placed on a hook in our laundry room. Over the weekend, I refill the water bottles and bag with all the items listed in the recipe. If I do not have the recipe ingredients, I add them to my weekly grocery list.

Laura Lee made the suggestion of having a "use by" sticker on the front of the bag so that we would know when we should dis-card the meal. I liked the idea, because not everyone will choose to have 365 breakfast and dinner bag meals and eat them twice a day as we do. Therefore, for those who do not plan on eating all of their bag meals in a year, some bag meals could theoretically expire before they are eaten.

One of the food storage problems I have faced over the years is a family member pulling items from the back of the storage shelf instead of the front. This ruins your food storage rotation scheme because the newly purchased items are being eaten before the older front facing ones.

Because of my problems with food storage rotation in the past, I developed a color-coded "use by" system that even a young child can understand. Here is how it works: for a given year of expiration there is a specific color of label. For instance, a bag meal that will expire in 2010 might have a green label next to the recipe sleeve. Not only do I put the month and year that the item is best used by, but I also have the label color coded for those who can't read yet. On the door of our food storage room I have a single sheet of paper that shows the color of the label that may be pulled for breakfast or dinner. For example, I might tell my grandson to go down to the basement and pick any bag meal he would like to eat that has a green sticker on the front.

10. You are always prepared to share a meal at a moment's notice.

Countless times in the past, I heard that a friend was having a difficult day or perhaps had just come home from having surgery or having a baby, and I wanted to bring dinner to her but had to wait because I didn't have enough advanced notice to grocery shop for the meal or even make an additional meal from what I had in the house. Bag meals have solved this problem for me. Now, I simply grab a bag meal off of the shelf, and within twenty minutes, I have a complete meal for a friend and her family. By the way, bag meals also make wonderful gifts for bridal kitchen showers!

11. In times of disaster, you won't be eating a granola bar and beef jerky. You will be enjoying a complete meal.

In a typical bag meal you can fit all of the entrée ingredients, cooking water, and a can of fruit and vegetables.

12. Suppose you are asked to evacuate your home. Adults can comfortably carry a bag meal in each hand, and even a young child can carry a bag meal in their arms.

We have a 72-hour emergency backpack ready for each member of our family, and next to the backpacks, we have enough bag meals for each person to carry two. This means that with the food in our 72-hour backpacks and our bag meals, we will have more than a week's worth of food and water for our family to eat and to share with others.

Contents of bag meal

Complete bag meal ready for pantry

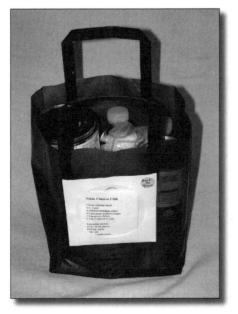

Bag meal bag with CD sleeve

Recipe with "use by" sticker

Storage shelves: 16 feet by 4 feet, containing 422 bag meals

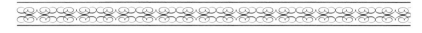

Bag Meal
Dinner Recipes

Casablanca Chicken

3 cups water

RESEALABLE BAG:
  2 cups rice

RESEALABLE BAG:
  2 Tbsp. dried bell pepper
  ½ cup raisins
  ½ cup dried apricots
  1 Tbsp. dried onion
  8 dehydrated tomato slices, cut into strips
  ½ tsp. garlic powder
  1 tsp. turmeric
  ¼ tsp. dried red pepper flakes
  1 Tbsp. brown sugar

PANTRY ITEMS:
  4 Tbsp. peanut butter
  1 Tbsp. balsamic vinegar

1 (13-oz.) can chicken meat, including broth
  or 1 pint home-canned chicken breast
1 (14.5-oz.) can diced Mexican tomatoes in
  lime juice and cilantro
1 (15-oz.) can garbanzo beans, including juice
1 (8-oz.) can tomato sauce

IN A HEAVY BOTTOMED SAUCEPAN with a tight fitting lid, on high heat, boil 3 cups water and 2 cups rice for 15 minutes. Remove from heat and set aside for 10 minutes then fluff. Do not lift the lid and peek at the rice while it is resting. If you have a rice cooker, just add water and rice and depress the lever. Fluff rice when lever pops up.

SIMMER REMAINING INGREDIENTS in a large saucepan over medium heat until dried fruits and vegetables are tender, about 15 minutes. To serve, place cooked rice on plates and top with Casablanca chicken sauce.

Santa Fe Chicken Pilaf

(Original recipe from the kitchen of Belinda Craft)

1 (3.8-oz.) package rice pilaf mix
2 cups water

PANTRY ITEM:
1 chicken bouillon cube

1 (14.5-oz.) can Mexican diced stewed tomatoes
1 (13-oz.) can chicken meat, drained or 1 pint home-canned chicken breast
1 (15-oz.) can corn, drained
1 (15-oz.) can black beans, drained and rinsed.
*Diced avocado, fresh cilantro sprigs

IN A MEDIUM SAUCEPAN, prepare rice pilaf according to directions on package, using water and bouillon. When rice pilaf is finished cooking, add remaining ingredients. Simmer until thoroughly heated.

*IN TIMES OF PLENTY, avocado and fresh cilantro make a nice addition.

Enchilada Soup

1 (1.5-oz.) enchilada sauce packet
1 (8-oz.) can tomato sauce
4 cups water
1 chicken bouillon cube
1 (12-oz.) can evaporated milk
1 (14.5-oz.) can diced tomatoes
1 (15-oz.) can black beans, drained
1 (13-oz.) can chicken meat, drained or 1 pint
   home-canned chicken breast

Garnish:

*Diced avocado, tortilla chips, olives, and fresh
   cilantro sprigs

In a large saucepan over medium heat, simmer all of the ingredients until hot. Garnish as desired.

*In times of plenty, avocado, tortilla chips, olives, and fresh cilantro make nice additions.

Hawaiian Haystacks

1 cup water
1 (13-oz.) can chicken meat, including broth
   or 1 pint home-canned chicken breast
1 (10.75-oz.) can condensed cream of
   chicken soup, do not reconstitute
2 (6-oz.) bags Chinese rice noodles

RESEALABLE BAG:
   1 chicken bouillon cube
   1½ tsp. dried basil
   1 tsp. garlic powder
   ¼ tsp. black pepper
   2 Tbsp. dried bell peppers
   8 dried tomato slices, cut in ¼-inch strips

RESEALABLE BAG
   3 Tbsp. cornstarch

TOPPINGS:
   4 Tbsp. sliced almonds
   1 (15-oz.) can mandarin oranges, drained
   1 (20-oz.) can pineapple, drained
   8 Tbsp. shredded sweetened coconut
   1 (8-oz.) canned cheddar cheese, shredded
   *Fresh diced tomatoes and green onions

IN A HEAVY SAUCEPAN over medium heat, combine ¾ cup water, chicken, cream of chicken soup, and the contents of first resealable bag. Mix cornstarch with remaining water and pour into saucepan. Stir until mixture thickens. To serve, place Chinese rice noodles on plate and spoon chicken sauce over noodles, followed with desired toppings.

*IN TIMES OF PLENTY, fresh tomato and green onions make nice additions.

Chicken Chow Mein

3½ cups water
1 (13-oz.) can chicken meat including broth
  or 1 pint home-canned chicken breast
1 (15-oz.) can bean sprouts, drained
2 (4-oz. cans sliced mushrooms, drained
1 (8-oz.) can bamboo shoots, drained
1 (15-oz.) can baby corn, drained
1 (8-oz.) can water chestnuts

RESEALABLE BAG:
  4 chicken bouillon cubes
  ⅓ cup dried onions
  ⅔ cup dried celery
  ⅔ cup dried carrot

RESEALABLE BAG:
  3 Tbsp. cornstarch

2 (6-oz.) bags Chinese rice noodles

IN A LARGE HEAVY SAUCEPAN over medium heat, combine 3 cups water and the next six ingredients. Add contents of first resealable bag, and heat thoroughly. Mix remaining water with cornstarch and add to saucepan, stirring constantly. When mixture has thickened to desired consistency, serve over Chinese rice noodles.

Mexican Chicken Soup

1 (13-oz.) can chicken meat, including broth
  or 1 pint home-canned chicken breast
2 (8-oz.) cans tomato sauce
1 (14.5-oz.) can Mexican diced tomatoes in
  lime juice and cilantro
1 (15-oz.) can black beans, including juice
1 (4-oz.) can diced green chilies
1 (15-oz.) can corn, including juice

RESEALABLE BAG:

½ tsp. ground cumin
½ tsp. powdered coriander
½ tsp. garlic powder
1 Tbsp. granulated white sugar

GARNISH:

*Diced avocado, sour cream, tortilla chips, olives,
  and fresh cilantro sprigs

IN A LARGE SAUCEPAN over medium heat, simmer all ingredients until thoroughly heated. Garnish as desired.

*IN TIMES OF PLENTY, avocado, sour cream, tortilla chips, olives, and fresh cilantro make nice additions.

Roast Beef Dinner

2 (12-oz.) cans roast beef or 1 quart home-
canned roast beef
1 (6.4-oz.) package instant potatoes, use
powder milk and water according to package directions
3 cups water
1 (12-oz.) bottle savory beef gravy
1 canned vegetable of choice

RESEALABLE BAG:
3 Tbsp. powdered milk

REMOVE ROAST BEEF FROM bottle and heat. Warm can of
vegetables. Prepare instant potatoes according to directions
on package, using powdered milk and water. Heat canned
gravy.

*THIS IS ONE OF our family's favorite Sunday meals. It is
amusing when dinner guests comment on how delicious the
roast is and want to know how I made it home from church
at 4:30 p.m. and had dinner ready by 5:00 p.m. Truly, heated
home-canned roast tastes like it was just taken out of the
oven. Add a can of vegetables, instant potatoes, and a bottle
of peaches, and you have a complete meal.

Sweet Roast with Tortillas

1¹/₃ cups water

RESEALABLE BAG:
  2 cups Masa Harina De Maiz mix

2 (12-oz.) cans roast beef or 1 quart home-
  canned roast beef
1 cup salsa

RESEALABLE BAG:
  1 cup brown sugar

GARNISH:
  *Avocado, fresh cilantro, tomatoes, sour cream,
  and black olives

IN A MEDIUM BOWL, combine water and tortilla mix. Knead dough about 5 minutes. Shape tortilla dough into 1½-inch balls. Press balls of tortilla dough between wax paper using a rolling pin or a tortilla press. Remove wax paper and cook each side of the tortilla 30 to 45 seconds on a hot griddle. Tortillas should be soft and pliable. Stack tortillas and cover with a slightly damp cloth to keep warm and soft.

IN A LARGE SAUCEPAN over medium heat, combine roast beef, brown sugar, and salsa. To serve, place 3 to 4 spoonfuls of meat mixture in a warm tortilla. Garnish as desired.

*IN TIMES OF PLENTY, avocado, fresh cilantro, tomatoes, sour cream, and black olives make nice additions.

Sweet Roast with Tortillas

Sun-Dried Tomato Pasta

(Original recipe from the kitchen of Belinda Craft)

> 5 cups water
> 8 oz. penne pasta
> 1 (14.5-oz.) can Italian diced tomatoes
> 1 (12-oz.) can evaporated milk
> 1 (13-oz.) can chicken meat, including broth
> or 1 pint home-canned chicken breast

RESEALABLE BAG:
> 1 chicken bouillon cube
> 1½ Tbsp. dried basil
> ½ tsp. garlic powder

IN A LARGE POT, bring 5 cups water to boil. Add pasta to boiling water and stir gently. Return to boil and cook 6 to 7 minutes or according to desired pasta tenderness. Remove from heat and drain. In a medium saucepan, combine remaining ingredients. Cook over medium heat for 10 minutes. To serve, gently toss sun dried tomato sauce and pasta until pasta is thoroughly coated.

Clam Chowder

2 (15-oz.) cans diced potatoes, drained
5½ cups water
1 (12-oz.) can evaporated milk
2 (6.5-oz.) cans clams, including juice

RESEALABLE BAG:
  1 Tbsp. dried celery
  2 Tbsp. dried carrot
  2 Tbsp. dried onions
  1 Tbsp. dried parsley
  ½ tsp. celery salt
  ¼ tsp. black pepper
  2 chicken bouillon cubes

RESEALABLE BAG:
  2 Tbsp. cornstarch

COMBINE IN A LARGE saucepan all ingredients except cornstarch and ½ cup water. Cook over medium heat, stirring frequently, for 15–20 minutes or until vegetables are tender. In a small bowl, combine cornstarch and remaining water. Stir until smooth. While stirring chowder, pour cornstarch solution into saucepan. Continue stirring until chowder is thick.

Turkey Dinner

(Original recipe from the kitchen of Belinda Craft)

1 (8-oz.) stuffing mix
1 cup water
1 (10.75-oz.) can cream of celery soup
2 (13-oz.) cans turkey meat, drained or 1
    quart home-canned turkey breast
1 (15-oz.) can green beans, drained
1 (12-oz.) can evaporated milk

RESEALABLE BAG:
1 Tbsp. dried onion
¼ tsp. powdered sage
⅛ tsp. black pepper

1 (1.25-oz.) envelope turkey gravy mix

PREHEAT OVEN TO 350°F. Mix together stuffing and water. Press ⅔ of the mixture into a 9" x 9" pan. In a medium saucepan over medium heat, mix together soup, meat, drained vegetables, ½ cup evaporated milk, and contents of resealable bag. Heat thoroughly and pour over stuffing. Spread remaining stuffing over top. Bake entrée for 30 minutes.

IN A SMALL SAUCEPAN over medium heat, prepare gravy using package directions and remaining evaporated milk, adding water if needed. To serve, place turkey entrée on plate and top with gravy.

Shepherd's Pie

1 (6.4-oz.) package instant potatoes
3 cups water

RESEALABLE BAG:
　3 Tbsp. powdered milk

1 (15-oz.) can green beans, drained
1 (10.75-oz.) can condensed tomato soup, do
　not reconstitute
1 (15-oz.) can corn, drained
1 (12-oz.) can roast beef or 1 pint home-
　canned hamburger
1 (8-oz.) canned cheddar cheese, shredded

PREHEAT OVEN TO 350°F. Prepare instant potatoes according to package directions using water and powdered milk. Mix green beans, tomato soup, corn, and hamburger in a 2½-quart casserole dish. Spread prepared potatoes over the mixture. Top with shredded cheddar cheese. Bake uncovered for 25 minutes.

Garbanzo Beans in Coconut Milk

3 cups water

RESEALABLE BAG:
  2 cups rice

2 (15-oz.) cans garbanzo beans*, including
  juice
1 (13.5-oz.) can coconut milk
1 (14.5-oz.) can chopped tomatoes with lime
  juice and cilantro

RESEALABLE BAG:
  ½ tsp. garam masala
  ½ tsp. powdered coriander
  1 tsp. garlic powder
  1 tsp. turmeric
  2 tsp. curry powder
  1 Tbsp. granulated white sugar
  ¼ tsp. cayenne pepper

IN A HEAVY BOTTOMED saucepan with a tight fitting lid, combine 3 cups water and 2 cups rice. Boil on high heat for 15 minutes. Remove from heat and set aside for 10 minutes, then fluff. Do not lift the lid while rice is resting. If you have a rice cooker, just add water and rice and depress the lever. Fluff rice when lever pops up.

IN A LARGE SAUCEPAN, combine garbanzo beans, coconut milk, tomatoes, and contents of the remaining resealable bag. Simmer over medium heat for 10 minutes. To serve, spoon garbanzo bean mixture over a serving of rice.

*GARBANZO BEANS AND CHICKPEAS are different names for the same bean.

Chicken Cacciatore

6 cups water
1 (12-oz.) package fettuccine pasta
2 (13-oz.) cans chicken meat, drained or 1 quart
    home-canned chicken breast
2 (14.5-oz.) cans diced Italian tomatoes
2 (8-oz.) cans tomato sauce
½ cup Concord grape juice

RESEALABLE BAG:
6 Tbsp. dried onions
1 tsp. garlic powder
½ tsp. black pepper
1 tsp. whole celery seed
1½ tsp. dried oregano
3 bay leaves

RESEALABLE BAG:
½ cup Parmesan cheese

IN A LARGE POT, bring water to a boil. Add pasta to boiling water and stir gently. Cook 6 to 7 minutes or according to desired pasta tenderness. Remove from heat and drain.

IN A LARGE SAUCEPAN, combine all remaining ingredients except Parmesan cheese, and simmer over medium heat for 20 minutes.

TO SERVE, TOP INDIVIDUAL servings of pasta with cacciatore sauce and sprinkle with Parmesan cheese.

Chili

1 (14.5-oz.) can diced tomatoes

1 (15-oz.) can red kidney beans, including juice

1 (15-oz.) can chili beans, including juice

1 (15-oz.) can black beans, including juice

1 (10.75-oz.) can condensed tomato soup, do not reconstitute

1 (12-oz.) can roast beef or 1 pint home-canned hamburger

RESEALABLE BAG:

2 Tbsp. dried onions

3 Tbsp. granulated white sugar

1 Tbsp. chili powder

1 tsp. ground cumin

½ tsp. ground cinnamon

¼-½ tsp. cayenne powder, according to taste

MIX ALL INGREDIENTS IN a large saucepan and cook on medium heat until thoroughly heated.

Chili

Potato Chicken Bake

1 (10.75-oz.) can condensed cream of
chicken soup, do not reconstitute

RESEALABLE BAG:
6 Tbsp. powdered milk

2½ cups water
1 (13-oz.) can chicken meat, drained or 1
pint home-canned chicken breast
1 (5-oz.) box sour cream and chive potatoes
or potato au-gratin

PREHEAT OVEN TO 350°F. In a medium bowl, mix soup, milk, and water. Add drained chicken and box of potatoes. Transfer to a greased 8" x 8" pan. Bake uncovered for 50 to 60 minutes, or until potatoes are tender.

Chicken Tetrazzini

(Original recipe from the kitchen of Belinda Craft)

6 cups water

RESEALABLE BAG:
8-oz. angel hair pasta, broken into thirds

PANTRY ITEMS:
8 Tbsp. canned butter

RESEALABLE BAG:
½ cup flour
1 tsp. salt
½ tsp. garlic powder
½ tsp. black pepper
2 chicken bouillon cubes

1 (12-oz.) can evaporated milk

RESEALABLE BAG:
½ cup Parmesan cheese

1 (4-oz.) can sliced mushrooms, drained
2 (13-oz.) cans chicken meat, drained or 1
quart home-canned chicken breast

BRING 4 CUPS WATER to a boil. Add pasta to boiling water and stir gently. Boil for 6 to 7 minutes or according to desired tenderness. Remove from heat and drain.

IN A MEDIUM SAUCEPAN over medium heat, melt butter and add spice packet, milk, remaining water, and Parmesan cheese. Cook, stirring frequently, until sauce thickens. Stir in mushrooms and chicken meat. To serve, place pasta on plate and top with tetrazzini sauce.

Ham Fried Rice

3 cups water

RESEALABLE BAG:
  2 cups rice

RESEALABLE BAG:
  2 Tbsp. dried chives, rehydrated
  3 Tbsp. dried celery, rehydrated

1 (16-oz.) canned ham, diced
1 (8-oz.) can water chestnuts, drained and
  chopped
1 (4-oz.) can sliced mushrooms, drained
1 (15-oz.) can peas and carrots, drained

RESEALABLE BAG:
  2 Tbsp. dried chives, rehydrated
  3 Tbsp. dried celery, rehydrated

PANTRY ITEMS:
  2 Tbsp. soy sauce
  2 Tbsp. vegetable oil
*2 scrambled eggs, cooked

TO COOK RICE: IN a heavy bottomed saucepan with a tight fitting lid, combine 3 cups water and 2 cups rice. Boil on high heat for 15 minutes. Remove from heat and set aside for 10 minutes, then fluff. Do not lift the lid while rice is resting. If you have a rice cooker, just add water and rice and depress the lever. Fluff rice when lever pops up.

REHYDRATE CELERY AND CHIVES by placing them in a small bowl along with 10 tablespoons with liquid from any canned item. It takes 5 to 10 minutes to rehydrate. In a large pan, heat oil over a medium-high heat. In a bowl, mix all ingredients together. Once oil is heated, fry ingredients until thoroughly heated.

*IF EGGS ARE AVAILABLE, scramble and toss into fried rice before serving.

Dahl Soup

4 cups water

RESEALABLE BAG:
  4 chicken bouillon cubes
  2 cups red lentils
  ½ tsp. garlic powder
  ½ cup dried onion
  ½ tsp. turmeric
  1 Tbsp. garam masala
  ¼ tsp. cayenne pepper
  1 tsp. ground cumin
  1 Tbsp. granulated white sugar
  *Fresh cilantro sprigs

1 (14.5-oz.) can diced tomatoes with lime
  and cilantro
1 (13.5-oz.) can coconut milk

IN A LARGE SAUCEPAN over medium heat combine all ingredients, except tomatoes and coconut milk. When lentils are tender, add tomatoes and coconut milk. Serve when soup is hot.

*IF AVAILABLE, GARNISH WITH fresh cilantro sprigs.

Salmon Patties

3 cups water

RESEALABLE BAG:
  2 cups rice

1 (14.75-oz.) can salmon or 1 pint home-
canned salmon

1 (4-oz.) can sliced mushrooms, drained

RESEALABLE BAG:
  1 Tbsp. dried onion, rehydrated
  12 soda crackers, crushed
  ½ tsp. chicken bouillon granules
  ¼ tsp. dried dill

PANTRY ITEMS:
  2 Tbsp. vegetable oil

1 (15-oz.) can beets, drained and halved

2 eggs, beaten

TO COOK RICE: IN a heavy bottomed saucepan with a tight fitting lid, combine 3 cups water and 2 cups rice. Boil on high heat for 15 minutes. Remove from heat and set aside for 10 minutes, then fluff. Do not lift the lid while rice is resting. If you have a rice cooker, just add water and rice and depress the lever. Fluff rice when lever pops up.

MIX ALL INGREDIENTS *EXCEPT* oil, rice, and beets, and shape mixture into patties. Place salmon patties in hot oiled pan and cook until brown. Flip and cook other side until brown. Serve with cooked rice or mashed potatoes and hot beets.

Salmon Patties

Mashed Potatoes with Chicken Gravy

(Original recipe from the kitchen of Robyn Hutchinson)

POTATOES:
1 (6.4-oz.) package instant potatoes

RESEALABLE BAG:
3 Tbsp. powdered milk

3²/₃ cups water

GRAVY:
1 (10.75-oz.) can condensed cream of chicken
    soup, do not reconstitute

2 cups water

2 chicken bouillon cubes

1 (12-oz.) can evaporated milk

8 Tbsp. canned butter

1 (13-oz.) can chicken meat, drained or 1
pint home-canned chicken breast

1 (15-oz.) can green beans, drained

PREPARE POTATOES ACCORDING TO package directions using milk and water.

COMBINE GRAVY INGREDIENTS IN a medium saucepan and heat thoroughly over medium heat. Add chicken to gravy. To serve, place potatoes on plate and top with gravy. Heat green beans and serve on the side.

NOTE: THE BRAND OF instant potatoes I use requires 2²/₃ cups boiling water and 1 cup milk; hence, my bag meals have 1 cup water to reconstitute the 3 tablespoons of powdered milk for a total of 3²/₃ cups water.

White Chicken Chili

2 (13-oz.) cans chicken meat, drained or 1
  quart home-canned chicken breast
4 cups water
2 (15-oz.) cans great northern beans, drained
1 (4-oz.) can diced green chilies
1 (12-oz.) can evaporated milk

RESEALABLE BAG:
  4 chicken bouillon cubes
  1/3 cup dried onions
  1½ tsp. garlic powder
  1 tsp. salt
  1 tsp. ground cumin
  1 tsp. oregano
  ½ tsp. black pepper
  ¼ tsp. cayenne pepper

*1 cup sour cream and 1 pint whipping
  cream

COMBINE ALL INGREDIENTS IN a large heavy saucepan and simmer over a medium heat. Serve hot.

*IF TIMES ARE PLENTIFUL, add 1 cup sour cream and 1 pint whipping cream for a richer chili.

Hamburger Pasta Casserole

6 cups water

8-oz. uncooked egg noodles

1 (12-oz.) can roast beef or 1 pint home-
canned hamburger

1 (10.75-oz.) can condensed tomato soup, do
not reconstitute

RESEALABLE BAG:

2 Tbsp. dried onion

1 (6-oz.) bottle chili sauce

1 tsp. chili powder

1/4 tsp. black pepper

1/2 cup canned cheddar cheese, shredded

PREHEAT OVEN TO 350°F.

IN A LARGE POT, bring water to a boil. Add pasta to boiling water and stir gently. Boil for 6 to 7 minutes or according to desired noodle tenderness. Remove from heat and drain.

IN A LARGE SAUCEPAN over medium heat, simmer all remaining ingredients *except* cheddar cheese. Combine noodles and hot hamburger chili sauce in a 2½ quart casserole dish and top with shredded cheddar cheese. Bake uncovered for 25 minutes or until cheese is melted.

Linguine in Clam Sauce

6 cups water
8-oz. uncooked linguine pasta

RESEALABLE BAG:
  2 Tbsp. dried onion
  1 Tbsp. flour
  1 tsp. dried basil
  1 Tbsp. dried parsley
  1 tsp. lemon pepper

2 Tbsp. canned butter, melted
2 (6.5-oz.) cans minced clams, reserve juice

IN A LARGE POT, bring water to a boil. Add pasta to boiling water and stir gently. Boil for 6 to 7 minutes according to desired pasta tenderness. Remove from heat and drain. In a small pan over medium heat, mix spice packet into melted butter and clam juice. Once flour is dissolved and there are no lumps, simmer for 5 minutes or until sauce is as thick as gravy. Add clams to gravy and cook 2 to 3 minutes. To serve, spoon clam sauce over a serving of linguine.

Meatless Chili

(Original recipe from the kitchen of Belinda Craft)

1 (15-oz.) can corn, including juice
2 (14.5-oz.) cans stewed tomatoes
2 (8-oz.) cans tomato sauce
1 (15-oz.) can black beans, including juice
1 (15-oz.) can red kidney beans, including juice

RESEALABLE BAG:
½ cup dried onion
1 cup dried green peppers
2 Tbsp. chili powder
¼ tsp. black pepper
1 Tbsp. granulated white sugar

COMBINE ALL INGREDIENTS IN a large saucepan and simmer over medium heat until hot.

Chicken Pasta Soup

8 cups water
1 (13-oz.) can chicken meat, including broth
  or 1 pint home-canned chicken breast
1 (12-oz.) can evaporated milk

RESEALABLE BAG:
  2 cups dried elbow pasta
  2 Tbsp. dried onion
  4 Tbsp. dried carrot
  2 Tbsp. dried celery
  ¼ tsp. garlic powder
  1 tsp. dried oregano
  ½ tsp. powdered thyme
  ½ tsp. poultry seasoning
  6 chicken bouillon cubes

IN A LARGE SAUCEPAN, over medium heat, mix together all ingredients. Be sure to stir frequently to prevent pasta from sticking to bottom of pan. Serve when soup reaches desired thickness and pasta is tender, about 20 minutes.

Beef Stroganoff

8 cups water

16-oz. uncooked egg noodles

1 (12-oz.) can roast beef or 1 pint home-
canned roast beef

1 (10.75-oz.) can condensed beefy mushroom soup, do not reconstitute

1 (10.75-oz.) can condensed cream of mushroom soup, do not reconstitute

1 (8-oz.) can sliced mushrooms, drained

1 (12-oz.) can evaporated milk

RESEALABLE BAG:
    3 Tbsp. dried onions
    1 beef bouillon cube

*1 cup sour cream

IN A LARGE POT, bring water to boil. Add pasta to boiling water and stir gently. Boil for 6 to 7 minutes according to desired noodle tenderness. Remove from heat and drain.

IN A LARGE SAUCEPAN, combine remaining ingredients* and simmer over medium heat. Toss noodles with beef stroganoff sauce.

*IF YOU HAVE SOUR cream, add 1 cup for extra richness.

Beef Stew

1 (12-oz.) can roast beef or 1 pint home-canned
   beef, reserve broth
2 (15-oz.) cans whole potatoes, drained and
   quartered
1 (15-oz.) can corn, drained
1 (15-oz.) can peas and carrots, drained
1 (12-oz.) jar savory beef gravy

RESEALABLE BAG:
2 Tbsp. dried onions
1 tsp. dried oregano

IN A LARGE SAUCEPAN mix all ingredients. Add beef broth from canned beef until desired consistency is obtained. Warm over medium heat and serve.

Salmon Dinner

1 (14.75-oz.) can salmon, drained or 1 pint
   home-canned salmon
1 (15-oz.) can vegetable of choice, drained
1 (6.4-oz.) package instant potatoes
3²/₃ cups water

RESEALABLE BAG:
3 Tbsp. powder milk
1 can fruit of choice

HEAT SALMON AND VEGETABLES separate pans or microwave in separate dishes. Prepare instant potatoes according to directions on package, using water and powder milk. Serve with fruit.

Spicy Jamaican Chicken and Rice

1 cup water

RESEALABLE BAG:
　1 cup rice

1 (13-oz.) can chicken meat, including broth
　or 1 pint home-canned chicken breast
1 (15-oz.) can black beans, drained and
　rinsed
1 (14.5-oz.) can diced tomatoes with lime
　juice and cilantro

RESEALABLE BAG:
　½ tsp. garlic powder
　2 Tbsp. dried onions
　1 tsp. curry powder
　1 tsp. thyme
　½ tsp. allspice
　½ tsp. crushed red pepper flakes
　½ tsp. black pepper

TO COOK RICE: IN a heavy-bottomed saucepan with a tight fitting lid, place 1 cup water and add 1 cup rice. Boil on high heat for 15 minutes. Remove from heat and set aside for 10 minutes, then fluff. Do not lift the lid while rice is resting. If you have a rice cooker, just add water and rice and depress the lever. Fluff rice when lever pops up.

IN A MEDIUM SAUCEPAN, combine remaining ingredients. Simmer over medium heat for 10 minutes. Spoon rice onto plate and top with chicken mixture to serve.

Hamburger Minestrone Soup

1 (12-oz.) can roast beef, drained or 1 pint
   home-canned hamburger
5 cups water
1 (10.75-oz.) can condensed tomato soup, do
   not reconstitute
2 (8-oz.) cans tomato sauce

RESEALABLE BAG:
   1 cup small shell macaroni
   1/3 cup dried onion
   1 tsp. garlic powder
   2 beef bouillon cubes
   1/3 cup dried celery
   1/2 cup dried carrot
   2 tsp. dried oregano
   1 tsp. dried basil

1 (15-oz.) can garbanzo beans, including
   juice
1 (15-oz.) can kidney beans, including juice

RESEALABLE BAG:
   1/2 cup Parmesan cheese

IN MEDIUM SAUCEPAN OVER medium heat, add all ingredients, except beans and Parmesan cheese. When vegetables are tender, add beans. Thoroughly heat all ingredients. Top each bowl of soup with 1 tablespoon Parmesan cheese.

Caribbean Chicken and Rice

5 cups water

RESEALABLE BAG:
  2 cups rice

RESEALABLE BAG:
  2 tsp. garlic powder
  1 tsp. onion powder
  ½ tsp. black pepper
  4 Tbsp. brown sugar
  2 Tbsp. dried celery
  ¼ cup dried bell peppers
  1 Tbsp. dried parsley
  1 tsp. thyme
  1 tsp. curry powder
  1 Tbsp. chives
  2 Tbsp. dried onion
  2 chicken bouillon cubes

1 (15-oz.) can black-eyed peas, drained
1 (13-oz.) can chicken meat, including broth
  or 1 pint home-canned chicken breast

TO COOK RICE: IN a heavy bottomed saucepan with a tight fitting lid, place 3 cups water and add 2 cups rice. Boil on high heat for 15 minutes. Remove from heat and set aside for 10 minutes, then fluff. Do not lift the lid while rice is resting. If you have a rice cooker, just add water and rice and depress the lever. Fluff rice when lever pops up.

IN A MEDIUM SAUCEPAN, add spice packet to remaining water and cook over medium heat until peppers and celery are tender, about 10 minutes. Add black-eyed peas and chicken meat and heat thoroughly. To serve, pour black-eyed pea mixture over rice.

Caribbean Chicken and Rice

Saffron-Infused Paella

1 can chicken, roast beef, or shrimp or 1 pint
 home-canned chicken breast, roast beef, or
 spicy sausage*
3½ cups water

RESEALABLE BAG:
 2 cups rice

RESEALABLE BAG:
 2 chicken bouillon cubes
 3 Tbsp. dried bell pepper
 2 Tbsp. dried onion
 ½ tsp. salt
 1 tsp. garlic powder
 ¼ tsp. cayenne pepper
 1 bay leaf
 4–5 stigmas (strands) saffron

COMBINE ALL INGREDIENTS IN a large skillet with a tight fitting lid. Simmer for 25 to 30 minutes on medium heat or until rice is tender.

*WHEN I WANT A special dinner, I add chicken, shrimp, and sausage to the dish.

Shrimp Soup

2 (15-oz.) cans red kidney beans, drained

2 (14.5-oz.) cans diced tomatoes with lime and cilantro

2 (10-oz.) cans condensed cream of mushroom soup, do not reconstitute

4 (4-oz.) cans tiny shrimp

1½ cups water

RESEALABLE BAG:

2 tsp. garlic powder

2 Tbsp. dried chopped celery

2 Tbsp. dried onion

2 tsp. dried dill

MIX ALL INGREDIENTS TOGETHER in a medium pot and simmer for 10 minutes over medium heat.

Shrimp Linguine

8 cups water

1 (8-oz.) package linguine pasta

2 (8-oz.) cans tiny shrimp, including juice

1 (12-oz.) can evaporated milk

RESEALABLE BAG:

2 tsp. dried basil

1 tsp. garlic powder

2 chicken bouillon cubes

¼ cup all-purpose flour

PANTRY ITEM:

4 Tbsp. lemon juice

IN A LARGE POT, bring 7 cups water to a boil. Add pasta to boiling water and stir gently. Boil for 6 to 7 minutes or according to desired pasta tenderness. Remove from heat and drain.

COMBINE REMAINING WATER AND flour. Stir until mixture is smooth. Add remaining ingredients in a medium saucepan. Cook over medium heat, stirring frequently until thick and bubbly. Toss pasta with shrimp sauce.

Spaghetti

4 cups water
1 (8-oz.) package angel hair pasta
1 pint home-canned hamburger
1 (10-oz.) jar spaghetti sauce

RESEALABLE BAG:
  ½ tsp. garlic powder
  2 tsp. Italian seasoning
  ½ tsp. onion powder

RESEALABLE BAG:
  ½ cup Parmesan cheese

IN A LARGE POT over a high heat bring water to boil. Add pasta to boiling water and stir gently. Boil for 6 to 7 minutes according to desired pasta tenderness. Remove from heat and drain.

IN A LARGE PAN, combine hamburger, spaghetti sauce, and spice packet. Simmer over medium heat. To serve, top spaghetti with sauce and sprinkle with Parmesan cheese.

Beefy Bean Casserole

1 (28-oz.) can baked beans of choice
1 (12-oz.) can roast beef, drained or 1 pint
   home-canned roast beef, chunked

RESEALABLE BAG:
   ¼ cup dried onion

   1 (15-oz.) can green beans, drained

COMBINE ALL INGREDIENTS IN a medium saucepan. Warm over a medium heat. Serve with green beans, wheat-germ corn bread muffins (recipe on page 122), and sprout salad.

Beef and Barley Soup

8 cups water
1 (12-oz.) can roast beef or 1 pint home-
   canned roast beef, chunked

RESEALABLE BAG:
   7 beef bouillon cubes
   ¼ cup dried celery
   ½ cup dried carrot
   2 Tbsp. dried onion
   ½ cup barley
   ⅛ tsp. black pepper

COMBINE ALL INGREDIENTS IN a large saucepan with a tight fitting lid. Simmer over a medium heat for 25 minutes.

Swedish Beef over Rice

3½ cups water

RESEALABLE BAG:
 2 cups rice

1 (10.75-oz.) can condensed golden mushroom soup, do not reconstitute

PANTRY:
 1½ tsp. Worcestershire sauce
 4 Tbsp. canned butter

1 (12-oz.) can evaporated milk

RESEALABLE BAG:
 ¼ cup all-purpose flour

1 (12-oz.) can roast beef, drained

TO COOK RICE: IN a heavy bottomed saucepan with a tight fitting lid, place 3 cups water and add 2 cups rice. Boil on high heat for 15 minutes. Remove from heat and set aside for 10 minutes, then fluff. Do not lift the lid while rice is resting. If you have a rice cooker, just add water and rice and depress the lever. Fluff rice when lever pops up.

IN A MEDIUM SAUCEPAN, combine golden mushroom soup, Worcestershire sauce, butter, and evaporated milk. Stir until smooth. In a cup, mix flour with remaining water, stirring until smooth. Add flour mixture to mushroom soup sauce, stirring over medium heat until bubbly. Add roast beef and cook until thoroughly heated. Spoon Swedish beef sauce on top of rice and serve.

Pad Thai

6 cups water

1 (12-oz.) package rice vermicelli or rice stick noodles

1 (13-oz.) can chicken meat, drained or 1 pint home-canned chicken breast

PANTRY ITEM:

Pad Thai sauce

RESEALABLE BAG:

½ cup chopped peanuts

Fresh Cilantro

IN A LARGE SAUCEPAN over high heat, bring water to a rolling boil. Place noodles in boiling water and cook for 5 to 8 minutes. Drain noodles and rinse with cool water. Gently toss chicken meat and noodles with just enough Pad Thai sauce to coat noodles. Transfer noodles to serving platter. Top with chopped peanuts and fresh cilantro, if available.

Pad Thai

Moroccan Chicken Couscous

RESEALABLE BAG:

1 cup couscous

1¼ cups boiling water
1 (15-oz.) can garbanzo beans, drained
1 (13-oz.) can chicken, including broth or 1
   pint home-canned chicken breasts
1 (14.5-oz.) can diced tomatoes with lime
   juice and cilantro

RESEALABLE BAG:

1 tsp. powdered cilantro leaves
2 Tbsp. brown sugar
1 tsp. ground cumin
1½ tsp. cardamom powder
1 tsp. powdered coriander seeds
1 tsp. ground ginger
1 tsp. turmeric
1 tsp. ground cinnamon
⅛ tsp. hot pepper flakes
1 tsp. garlic powder
2 Tbsp. dried onion
2 Tbsp. dried chives

PLACE COUSCOUS IN A medium-sized glass bowl and cover with boiling water. Cover bowl with plastic wrap and set aside for 15 minutes. Combine remaining ingredients in a medium saucepan and simmer over a medium heat for 15 minutes. To serve, place couscous on a platter and top with Moroccan chicken sauce.

Chicken Pot Pie

FILLING:

1 (8-oz.) can water chestnuts, drained and
  chopped
1 (15-oz.) can mixed vegetables, including juice
1 (10.75-oz.) can condensed cream of chicken
  soup
1 cup water
1 (13-oz.) can chicken, drained or 1 pint home-
  canned chicken breast

RESEALABLE BAG:

1 Tbsp. dried onions
1 Tbsp. dried celery
½ tsp. poultry seasoning
3 Tbsp. of powdered milk

PREHEAT OVEN TO 350°F. Combine all ingredients in a 2½ quart casserole dish.

PIE CRUST:

PANTRY ITEM:

1 cup shortening

RESEALABLE BAG:

6 Tbsp. powdered milk
⅓ cup + 3 Tbsp. cold water

RESEALABLE BAG:

2 cups flour
½ tsp. salt

IN A MEDIUM BOWL, whip shortening, powdered milk, and water until fluffy. Using two knives or a pastry blender, cut in flour and salt. Roll out into desired shape. There will be extra dough for decorating the crust. Cover filling with pie crust. Bake uncovered for 30 minutes or until crust is golden brown.

Thai Chicken Soup

2 (14.5-oz.) cans fire roasted diced tomatoes
2 (13-oz.) cans chicken meat, including broth
   or 1 quart home-canned chicken breast
1 (13.5-oz.) can coconut milk

RESEALABLE BAG:
   2 Tbsp. dried onion
   1 tsp. garam masala
   1 tsp. powdered ginger
   1 tsp. garlic powder
   3 chicken bouillon cubes

PUREE FIRE ROASTED TOMATOES in a blender or food processor. In a large saucepan, combine all ingredients, including tomato puree and simmer over medium heat for 20 minutes.

Hardy Mac 'n' Cheese Soup

1 (12-oz.) can roast beef, drained or 1 pint
    home-canned hamburger
1 (10.75-oz.) can condensed cheddar cheese
    soup, do not reconstitute
2 (12-oz.) cans evaporated milk
2 cups water

RESEALABLE BAG:
    1½ cups uncooked elbow macaroni
    4 Tbsp. dried onion
    2 tsp. chili powder
    ¼ tsp. cayenne pepper
    2 tsp. ground cumin

PANTRY ITEMS:
    salsa
    1 bag corn tortilla chips

PLACE ALL INGREDIENTS, EXCEPT salsa and chips, in a large saucepan. Simmer over medium heat for 15 minutes, stirring frequently to prevent noodles from sticking to bottom of pan. Remove saucepan from heat; let stand for 5 minutes. Garnish with salsa and tortilla chips.

Red Lentil and Sausage Soup

8 cups water

RESEALABLE BAG:
  5 chicken bouillon cubes
  2 Tbsp. dried onion
  1 tsp. garlic powder
  2 tsp. ground cumin
  1 cup red lentils
  ¼ tsp. cayenne pepper

½ cup canned spinach*, drained
1 pint home-canned sausage

IN A LARGE SAUCEPAN over medium heat combine all ingredients, except spinach and sausage. Bring to a boil. Reduce heat to medium-low and simmer 25 minutes. Add spinach and sausage. Serve when sausage is thoroughly heated.

* **IN THE SUMMER,** we substitute fresh spinach from our garden.

Tacos

1 (12-oz.) can roast beef, drained or 1 pint home-
    canned hamburger
1 cup salsa
1 (8-oz.) can cheddar cheese
1 package taco shells, 12-count

IN A SKILLET OVER medium heat, warm hamburger. Build taco with hamburger, salsa, and grated cheddar cheese. I usually serve this entrée with Spanish rice (see below) and pinto beans (see page 68).

Spanish Rice

RESEALABLE BAG:
  2 cups rice

4 cups water

RESEALABLE BAG:
  4 chicken bouillon cubes
  1 tsp. garlic powder
  4 Tbsp. dried onion
  4 Tbsp. dried bell peppers

1 (8-oz.) can tomato sauce

RESEALABLE BAG:
  2 cups rice

IN A LARGE PAN, brown uncooked rice over medium-high heat. Add remaining ingredients, cover pan with a tight fitting lid, and cook for 20 to 25 minutes over medium-low heat.

Pinto Beans

(Original recipe from the kitchen of Laura Lee Andersen)

2½ cups water

RESEALABLE BAG:
1 tsp. salt
1 tsp. chicken bouillon
½ tsp. ground cumin
¾ tsp. chili powder
4 Tbsp. dried onions

RESEALABLE BAG
1 cup pinto bean flour*

IN A SMALL PAN, bring all ingredients, except pinto bean flour, to a boil. Stir in pinto bean flour. Cook for 5 minutes at a medium heat until thick.

*MOST WHEAT GRINDERS OR mills can safely mill pinto beans into flour; however, it is always a wise practice to consult the manufacturer or instruction manual before milling products for the first time to ensure safety and mill longevity.

"Night-Before" Lasagna*

BÉCHAMEL SAUCE

    5 Tbsp. canned butter
    4 cups water

RESEALABLE BAG:
    4 Tbsp. all-purpose flour
    12 Tbsp. powdered milk
    ½ tsp. nutmeg
    1 tsp. garlic powder
    1 tsp. dried basil
    1 tsp. oregano

10 lasagna noodles
1 (10-oz.) bottle spaghetti sauce
2 (12-oz.) cans roast beef, shredded and
    drained or 1 quart home-canned hamburger
1 (8-oz.) can cheddar cheese, shredded

PANTRY:
    1 cup Parmesan cheese

IN A MEDIUM SAUCEPAN, melt butter over medium-low heat. Add approximately ¼ resealable bag contents to melted butter until a paste is formed. Add water (about ¾ to 1 cup at a time) until incorporated. Continue alternating dry ingredients and water until both are incorporated. Bring mixture to a boil. Cook 10 minutes, stirring constantly, then remove from heat.

SPOON ENOUGH SPAGHETTI SAUCE to coat bottom of a 13" x 9" pan ¼-inch thick. Then, in order, layer half of lasagna noodles, hamburger, béchamel sauce, and Parmesan cheese. Top with remaining noodles, spaghetti sauce, and grated cheddar cheese. Cover with foil and refrigerate over night. The next day, bake at 350°F, uncovered for 25 minutes or until hot and bubbly.

*IF YOU FORGET TO make the lasagna the night before, just boil noodles until soft and follow recipe.

Savory Pancakes with Chicken & Plum Sauce

2 cups warm water

RESEALABLE BAG:
2 chicken bouillon cubes
2 Tbsp. dried celery
3 Tbsp. dried bell peppers
1 Tbsp. dried onion
1½ tsp. dried dill
¼ tsp. turmeric
1 tsp. granulated sugar
6 Tbsp. powdered milk

RESEALABLE BAG:
1¾ cups all-purpose flour

2 (13-oz.) cans chicken meat, drained or 1 quart home-canned chicken breast
1 (8-oz.) can sliced water chestnuts, drained
1 (8-oz.) can bamboo shoots, drained
1 (14-oz.) can bean sprouts, drained

PANTRY ITEM:
½ cup plum sauce

RESEALABLE BAG:
½ cup chopped peanuts
Green onions and cilantro, chopped*

IN A LARGE BOWL, combine water and contents of first resealable bag, and allow vegetables to rest for 10 minutes or until soft. Stir in flour. Ladle pancake batter onto a hot oiled griddle. Turn pancakes when edges are brown. While pancakes are cooking, in a small saucepan over medium heat, combine chicken, water chestnuts, bamboo shoots, bean sprouts, and enough plum sauce to coat ingredients. To serve, place hot pancake on plate and top with chicken plum sauce mixture, peanuts, green onions, and cilantro.

*IF YOU HAVE GREEN onions and cilantro from the garden, chop and sprinkle on top.

Chicken Korma

4 cups water

RESEALABLE BAG:
   2 cups rice

1 (13-oz.) can chicken meat, including broth
   or 1 pint home-canned chicken breast
2 cans tomato sauce

RESEALABLE BAG:
   1 tsp. ground cinnamon
   ½ tsp. ground cloves
   ½ tsp. cardamom powder
   4 Tbsp. dried onion
   1 tsp. garlic powder
   ½ tsp. crushed red pepper flakes
   ½ tsp. ground coriander
   ½ tsp. ground cumin
   1 Tbsp. sugar
   2 Tbsp. dried parsley

½ cup evaporated milk

TO COOK RICE: IN a heavy bottomed saucepan with a tight fitting lid, place 3 cups water and add 2 cups rice. Boil on high heat for 15 minutes. Remove from heat and set aside for 10 minutes, then fluff. Do not lift the lid while rice is resting. If you have a rice cooker, just add water and rice and depress the lever. Fluff rice when lever pops up.

IN A LARGE SAUCEPAN, combine remaining water and all remaining ingredients *except* evaporated milk. Cook for 10 minutes over medium heat. Stir in evaporated milk. To serve, top rice with chicken korma sauce.

Corn Flour Enchiladas in Green Sauce

1¼ cups water

RESEALABLE BAG:
2 cups Masa Tortilla Mix

FILLING:
¾ cup water
1 (12-oz.) can roast beef, shredded and
    drained or 1 pint home-canned hamburger
1 (15-oz.) can diced potatoes, drained
1 (15-oz.) can corn, drained
1 (4-oz.) can green chilies
2 (28-oz.) cans green enchilada sauce
1 cup canned cheddar cheese, shredded

IN A MEDIUM BOWL, combine tortilla mix and water. Knead dough about 5 minutes. Shape dough into 1½-inch balls. Press balls of tortilla dough between wax paper using a rolling pin. Remove from wax paper and cook each side of the tortilla 30 seconds on a hot griddle. Tortillas should be soft and pliable. Stack tortillas and cover with a cloth to keep warm.

PREHEAT OVEN TO 350° F. In a large saucepan over medium heat, combine first 4 ingredients and 1½ cups green enchilada sauce. Lightly grease a 13" x 9" baking dish. Spread 1 cup enchilada sauce on bottom of baking dish. Place 5 tablespoons of filling in the middle of tortilla and roll. Place tortilla in baking dish with overlapped side of tortilla facing the bottom of the baking dish; continue until baking dish is full. Pour remaining enchilada sauce on top of filled tortillas. Top with shredded cheddar cheese. Bake enchiladas for 25 minutes or until hot and bubbly.

Flour Tortilla Enchiladas in Red Sauce

2 cups water

RESEALABLE BAG:
  4 cups all-purpose flour
  1 tsp. salt

PANTRY ITEMS:
  1 tsp. yeast
  2 Tbsp. shortening
  ½ tsp. baking powder

FILLING:
  2 (1.5-oz.) enchilada sauce packets
  2 (12-oz.) cans chicken meat, drained or 1
    quart home-canned chicken breast
  1 15-oz. can black beans, drained
  1 cup canned cheddar cheese
  2 (8-oz.) cans tomato sauce
  3 cups water

RESEALABLE BAG:
  4 Tbsp. dried onion

DISSOLVE YEAST IN WARM water. In a separate bowl, combine dry ingredients. Mix in yeast. Let dough rise 15 minutes in an overturned bowl. Knead dough on floured surface for one minute, adding flour until dough is no longer sticky. Pinch off dough and roll or press into desired size. Cook on a hot griddle. Flip tortilla when bubbles develop.

PREHEAT OVEN TO 350°F. Prepare enchilada sauce according to package directions. Combine chicken, beans, and 1 cup enchilada sauce. Spread 1 cup of enchilada sauce on bottom of lightly greased 13" x 9" baking dish. Place 4 tablespoons of filling in the middle of tortilla and roll. Place tortilla in baking dish with overlapped side facing the bottom of the baking dish; continue until baking dish is full. Pour remaining enchilada sauce on top of filled tortillas. Top with shredded cheddar cheese. Bake enchiladas for 25 minutes or until hot and bubbly.

Chicken and Artichoke Couscous

1 (13-oz.) can or 1 pint home-canned chicken,
including juice
1 (13.5-oz.) can artichoke hearts, drained
1 (14.5-oz.) can diced tomatoes

RESEALABLE BAG:
3 Tbsp. dried onion
½ tsp. dried rosemary
1 tsp. powdered garlic
½ tsp. dried oregano
½ tsp. dried thyme
½ tsp. salt
¼ tsp. pepper
3 Tbsp. dried parsley

MIX ABOVE INGREDIENTS IN a large saucepan. Simmer over medium-low heat for 20 minutes.

COUSCOUS:

RESEALABLE BAG:
2 cups couscous
2½ cups boiling water

PLACE COUSCOUS IN A medium bowl. Pour boiling water over couscous and cover with a plate or plastic wrap. After 10 minutes, remove cover and fluff with a fork. Serve couscous with chicken topping.

Yummy Yammy Casserole

3 cups water

RESEALABLE BAG:
  2 cups cooked rice

2 (13-oz.) cans chicken breast or 1 quart
  home-canned chicken, drained
1 (14-oz.) can yams, drained and cubed
1 (14-oz.) can asparagus, drained
1 (14-oz.) can artichoke hearts, drained
2 (10.5-oz.) cans condensed cream of
  chicken soup, do not reconstitute

RESEALABLE BAG:
  2 tsp. curry powder

PANTRY:
  1 cup mayonnaise
  1 cup Miracle Whip Salad Dressing
  3 Tbsp. lemon juice
  2 cups shredded canned cheese

TO COOK RICE: **IN** a heavy bottomed saucepan with a tight fitting lid, place 3 cups water and add 2 cups rice. Boil on high heat for 15 minutes. Remove from heat and set aside for 10 minutes, then fluff. Do not lift the lid while rice is resting. If you have a rice cooker, just add water and rice and depress the lever. Fluff rice when lever pops up

GREASE 9" x 13" pan. Cover bottom with cooked rice. Sprinkle chicken breast, yams, and asparagus on top of rice. In a separate bowl, mix remaining ingredients and pour over chicken mixture. Bake at 350° for 25 minutes.

Veggie Soup with Hamburger

1 (12-oz.) can shredded roast beef, including
   broth or 1 pint home-canned roast beef or
   hamburger
2 cups water
1 (15-oz.) can corn, including juice
1 (14.5-oz.) can diced tomatoes
1 (15-oz.) can lima beans, drained
1 (14-oz.) can diced potatoes, drained

RESEALABLE BAG:
2 Tbsp. dried onion
¼ cup dried bell pepper
2 Tbsp. dried carrot
1 tsp. dried basil
4 beef bouillon cubes
1 bay leaf

PANTRY:
1 tsp. Worcestershire sauce

IN A LARGE POT, mix all ingredients and simmer over medium-low heat for 20 minutes.

Butternut Soup

2 cups water
1 (15-oz.) can butternut squash puree
1 (12-oz.) can evaporated milk

RESEALABLE BAG:
  2 chicken bouillon cubes
  ¼ cup dried carrots
  2 Tbsp. dried onion
  1 tsp. cinnamon

IN A MEDIUM POT, cook all ingredients over a medium-low heat for 20 minutes or until dried vegetables are tender and soup is warm.

Autumn Butter Bean Soup

5 cups water
2 (16-oz.) cans butter beans, including juice
1 (14-oz.) can diced tomatoes

RESEALABLE BAG:
  ½ cup dried celery
  ½ cup dried carrots
  4 Tbsp. dried chopped onion
  1 tsp. basil
  1 beef bouillon cube
  1 Tbsp. granulated white sugar
  1 ½ tsp. Spike seasoning
  ¼ tsp. pepper

PANTRY:
  1½ tsp. Worcestershire sauce

MIX ALL INGREDIENTS TOGETHER in a medium pot and simmer for 10 minutes over medium heat.

Uglier than Heck Soup

2 (12-oz.) cans roast beef, including broth or
   1 quart home-canned roast beef
1 cup Concord grape juice
2 1/2 cups water
6-oz. shell macaroni, uncooked
3 (14.5-oz.) cans stewed tomatoes
1 (15.5-oz.) can cannellini beans, drained
1 (15-oz.) can lima beans, drained
1 (15-oz.) can corn, drained

PANTRY:

  3 Tbsp. dried onion
  ¼ cup dried bell pepper
  1 tsp. salt
  ½ tsp. black pepper

MIX ALL INGREDIENTS IN a medium pot and simmer for 20 minutes over medium heat.

Spicy Chicken and Navy Bean Soup

2 (8-oz.) cans tomato sauce
1 (16-oz.) can navy beans, drained
1 (13-oz.) can chicken, drained or 1 pint
    home-canned chicken
4 cups water
1 (6.5-oz.) can sliced mushrooms

Resealable bag:
   3 Tbsp. dried onion
   1 tsp. garlic powder
   ¼ cup dried bell pepper
   ½ tsp. dried oregano
   ½ tsp. dried parsley
   ½ tsp. dried basil
   ¼ tsp. crushed red pepper flakes

Mix all ingredients together in a medium pot and simmer for 10 minutes over medium heat.

Bag Meal
Breakfast Recipes

Pancakes

PANCAKE BATTERS MAY REQUIRE more liquids due to the coarseness of the flour or batter thickening over time. Should the batter be too thick, just add water by the tablespoon until the batter is the desired consistency for pancake batter.

Coconut Pancakes, page 87

Whole-Wheat Pancakes with Evaporated Milk and Powdered Eggs

2 (12-oz.) cans evaporated milk

RESEALABLE BAG:
2½ cups wheat flour
3 Tbsp. powdered eggs
1 tsp. salt
2 tsp. baking powder
1 tsp. baking soda
3 Tbsp. granulated white sugar (optional)

PANTRY ITEM:
3 Tbsp. oil

IN A LARGE BOWL stir all ingredients until batter is smooth. Ladle pancake batter onto a hot oiled griddle. Turn pancakes when bubbles appear and edges are golden brown.

Whole Wheat Pancakes with Powdered Egg and Powdered Milk

3 cups water

RESEALABLE BAG:

3½ cups wheat flour
1 cup powdered milk
3 Tbsp. powdered eggs
1 tsp. salt
3 Tbsp. granulated white sugar
2 tsp. baking powder
1 tsp. baking soda

PANTRY ITEM:

3 Tbsp. oil

IN A LARGE BOWL stir all ingredients until batter is smooth. Ladle pancake batter onto a hot oiled griddle. Turn pancakes when bubbles appear and edges are golden brown.

Whole Wheat Pancakes with Fresh Buttermilk and Eggs*

(Original recipe from the kitchen of Rosalie Snow)

> 3 cups wheat flour
> 3 cups buttermilk
> 3 eggs
> 3 Tbsp. oil
> 2 tsp. baking powder
> 1 tsp. baking soda
> 1 tsp. salt
> 3 Tbsp. granulated white sugar

IN A LARGE BOWL stir all ingredients until batter is smooth. Ladle pancake batter onto a hot oiled griddle. Turn pancakes when bubbles appear and edges are golden brown.

*I INCLUDED THIS RECIPE because without fail, demonstration participants ask for a fresh buttermilk and egg pancake.

Coconut Pancakes

1 cup water

RESEALABLE BAG:
2 cups flour
½ cup granulated white sugar
½ tsp. salt
4 tsp. baking powder

RESEALABLE BAG:
1 cup sweetened coconut flakes

PANTRY ITEM:
1 tsp. vanilla extract

1 (13.5-oz.) can coconut milk

IN A LARGE BOWL stir all ingredients, except coconut flakes, until batter is smooth. Add coconut. Ladle pancake batter onto a hot oiled griddle. Turn pancakes when bubbles appear and edges are golden brown.

Chocolate Chip Pancakes

2¼ cups water

RESEALABLE BAG:
2 cups flour
¼ cup granulated white sugar
2 Tbsp. baking powder
1 tsp. salt
2 Tbsp. powdered eggs
5 Tbsp. powdered milk

PANTRY
1 tsp. vanilla

RESEALABLE BAG:
½ cup miniature chocolate chips

IN A LARGE BOWL add all ingredients, except chocolate chips. Stir until smooth. Add chocolate chips to batter. Ladle pancake batter onto a hot oiled griddle. Turn pancakes when bubbles appear and edges are golden brown.

Gingerbread Pancakes

1 cup water

RESEALABLE BAG:
  1 cup wheat flour
  ¼ tsp. salt
  3 Tbsp. brown sugar*
  ½ tsp. ground cinnamon
  ⅛ tsp. ground cloves
  ⅓ cup powdered milk
  1 Tbsp. powdered egg
  1 tsp. baking powder

PANTRY ITEM:
  1 Tbsp. oil

IN A LARGE BOWL stir all ingredients until batter is smooth. Ladle pancake batter onto a hot oiled griddle. Turn pancakes when bubbles appear and edges are golden brown.

*BE SURE TO COMBINE brown sugar with other ingredients when you bag your mix, or you will have a hard lump of brown sugar as time passes.

Cornmeal Pancakes

1 (12-oz.) can evaporated milk

RESEALABLE BAG:
- ½ cup flour
- 1 tsp. granulated white sugar
- 2 Tbsp. powdered eggs
- ¼ tsp. salt
- 1 cup corn meal
- 1 tsp. baking soda

IN A LARGE BOWL stir all ingredients until batter is smooth. Ladle pancake batter onto a hot oiled griddle. Turn pancakes when bubbles appear and edges are golden brown.

Pumpkin Chocolate Chip Pancakes

2 (15-oz.) cans evaporated milk
1 cup canned pumpkin puree

RESEALABLE BAG:
  2 cups wheat flour
  3 Tbsp. brown sugar*
  1 Tbsp. baking powder
  1¼ tsp. pumpkin pie spice
  1 tsp. salt
  1 Tbsp. powdered egg

RESEALABLE BAG:
  1 cup mini semi-sweet chocolate chips

PANTRY:
2 Tbsp. oil

IN A LARGE BOWL stir all ingredients, except for chocolate chips, until batter is smooth. Ladle pancake batter onto a hot oiled griddle. Turn pancakes when bubbles appear and edges are golden brown.

*BE SURE TO COMBINE brown sugar with other ingredients when you bag your mix or you will have a hard lump of brown sugar as time passes.

Gluten-Free Pancakes

1 (12-oz.) can evaporated milk
1 cup water

RESEALABLE BAG:
   1 cup rice flour
   1 cup buckwheat flour
   1 cup millet flour
   2 tsp. baking powder
   3 Tbsp. granulated white sugar
   1 tsp. salt
   2 Tbsp. powdered egg

PANTRY ITEMS:
   ¼ cup canola oil
   1 tsp. vanilla or almond extract

IN A LARGE BOWL stir all ingredients until batter is smooth. Ladle pancake batter onto a hot oiled griddle. Turn pancakes when bubbles appear and edges are golden brown.

Multi-Grain Pancakes

2 cups water

RESEALABLE BAG:

   6 Tbsp. powdered milk
   ½ cup oats
   ⅔ cup wheat flour
   ⅔ cup flax seed meal
   ¼ cup corn meal
   1½ tsp. baking powder
   ½ tsp. baking soda
   ¼ tsp. salt
   1 tsp. ground cinnamon
   2 Tbsp. powdered eggs
   ¼ cup brown sugar*

PANTRY:

   1 Tbsp. oil

IN A LARGE BOWL stir all ingredients until batter is smooth. Allow to rest for 5 minutes. Batter thickens during this time. Ladle pancake batter onto a hot oiled griddle. Turn pancakes when bubbles appear and edges are golden brown.

*BE SURE TO COMBINE brown sugar with other ingredients when you bag your mix or you will have a hard lump of brown sugar as time passes.

Oatmeal Blueberry Pancakes

2¼ cups water

RESEALABLE BAG:
- ½ cup wheat flour
- ½ cup all-purpose flour
- 4 Tbsp. brown sugar*
- 2 Tbsp. baking powder
- ½ tsp. salt
- 1½ cups oats
- 6 Tbsp. powdered milk
- 3 Tbsp. powdered eggs

PANTRY ITEM:
- ¼ cup oil

RESEALABLE BAG:
- ½–1 cup dried blueberries (For soft blueberries, cover the night before with ½ cup water. Drain in the morning and gently fold into batter.)

MIX ALL INGREDIENTS EXCEPT blueberries. Once batter is smooth, add blueberries. Allow to rest for 5 minutes. Ladle pancake batter onto a hot oiled griddle. Turn pancakes when bubbles appear and edges are golden brown.

*BE SURE TO COMBINE brown sugar with other ingredients when you bag your mix or you will have a hard lump of brown sugar as time passes.

Rye Pancakes

1 (12-oz.) can evaporated milk
⅓ cup + 2 Tbsp. water

RESEALABLE BAG:
   2¼ cups rye flour
   1½ tsp. baking powder
   1 tsp. baking soda
   1 tsp. salt
   ¾ cup brown sugar*
   4 Tbsp. powdered eggs

IN A LARGE BOWL stir all ingredients until batter is smooth. Ladle pancake batter onto a hot oiled griddle. Turn pancakes when bubbles appear and edges are golden brown.

*BE SURE TO COMBINE brown sugar with other ingredients when you bag your mix or you will have a hard lump of brown sugar as time passes.

Cinnamon Orange Pancakes

1 cup + 1 Tbsp. water
1 (6-oz.) single serving size can of orange
   juice

RESEALABLE BAG:

1 cup whole wheat flour
¾ cup all-purpose flour
3 Tbsp. powdered milk
1 Tbsp. powdered egg
2 Tbsp. wheat germ
2 Tbsp. granulated white sugar
2 tsp. baking powder
1 tsp. orange zest
1 tsp. ground cinnamon
¼ tsp. salt

IN A LARGE BOWL stir all ingredients until batter is smooth. Ladle pancake batter onto a hot oiled griddle. Turn pancakes when bubbles appear and edges are golden brown.

Pancake Syrups

MANY PANCAKE SYRUPS CAN be home-canned successfully. If you are going to home-can pancake syrups, bring syrup to boiling point and pour into sterilized bottles leaving a ½-inch head space. Place in boiling water bath with bottles submerged at least 2 inches below water level and process in water bath. Refer to your canner's processing booklet time-table for the correct time to process for your altitude.

MY FAMILY'S FAVORITE SYRUPS are buttermilk syrup, bottled jams that did not set, and pureed bottled fruit. I simply blend the fruit until smooth, and then warm and serve.

I PERSONALLY NEVER HOME-CAN any recipes that contain dairy products; however, there are individuals who believe that dairy products can be safely home-canned. If you are in doubt as to what can be safely home-canned, contact your state's agricultural extension.

IF YOU ARE GOING to store syrup for later use, remember to add a tablespoon or two of corn syrup to keep it from turning grainy over time.

Buttermilk Syrup

1 cup granulated white sugar
½ cup canned butter
½ cup buttermilk*
1 cup light corn syrup
1 tsp. baking soda
1 tsp. vanilla extract

IN A MEDIUM SAUCEPAN, bring all ingredients to a boil except vanilla extract. Boil for 7 minutes. Remove from burner and stir in vanilla extract. Stir until foam dissipates.

*I HAVE MADE THIS recipe with evaporated milk. The syrup was adequate.

Maple Syrup

2 cups granulated white sugar
1 cup water
½ tsp. maple flavoring

IN A MEDIUM SAUCEPAN over medium heat, bring sugar and water to a boil. Remove from heat and add maple flavoring.

Cinnamon Honey Syrup

1 cup honey
½ cup butter
1 tsp. ground cinnamon

COMBINE ALL INGREDIENTS in a small saucepan and simmer over medium heat until butter is melted.

Apple Cinnamon Syrup

1 cup granulated white sugar
2 Tbsp. cornstarch
2 cups apple cider or apple juice
2 tsp. ground cinnamon
1½ Tbsp. lemon juice
¼ cup butter

In a small saucepan, combine all ingredients except butter. Bring mixture to a boil over medium heat, stirring often. Remove saucepan from burner and add butter. Stir until butter is melted and well mixed.

Gelatin Pancake Syrup

1 (3-oz.) box flavored gelatin
½ cup granulated white sugar
2 Tbsp. cornstarch
1 cup water

Mix gelatin, sugar, cornstarch, and water together in a small saucepan. Bring mixture to a boil over medium heat, stirring frequently. Remove syrup from burner and cool. As syrup cools it will thicken.

Orange Honey Syrup

⅓ cup butter
⅓ cup honey
⅓ cup orange juice

Combine all ingredients in a small saucepan. Simmer 5 minutes over a medium heat. Serve warm.

Raspberry Syrup

¼ cup water
2 tsp. cornstarch
¾ cup frozen raspberry juice concentrate

MIX WATER AND CORNSTARCH together until smooth. Add juice. Bring ingredients to a boil.

Vanilla Syrup

6 Tbsp. water
¼ cup brown sugar
¼ cup granulated white sugar
1 tsp. vanilla extract

BRING ALL INGREDIENTS TO a boil in a small saucepan over medium heat.

Pineapple Syrup

½ cup granulated white sugar
½ cup butter
¼ cup frozen pineapple juice concentrate

IN A SMALL SAUCEPAN over medium heat, combine all ingredients and heat until hot but not to boiling point.

Coconut Syrup

1 cup corn syrup
1 cup granulated white sugar
6 tsp. powdered milk
1 cup water
1 tsp. coconut flavoring

IN A MEDIUM SAUCEPAN over medium heat, combine corn syrup and sugar. Stir until sugar is dissolved. Dissolve powdered milk into water. Remove from heat and stir in coconut flavoring.

Rich Coconut Syrup

½ cup canned butter
1 cup corn syrup
1 cup sugar
¼ cup water
1 Tbsp. coconut extract

MELT BUTTER IN MEDIUM saucepan over medium heat. Add remaining ingredients and stir until sugar is dissolved. Serve when syrup is hot.

Butter Maple Pancake Syrup

1 cup maple syrup
1 cup butter

IN A SMALL SAUCEPAN over medium heat stir ingredients until warm and butter is melted.

DON'T LAUGH WHEN YOU see this simple recipe! Breakfast guests frequently ask what the recipe is for my delicious syrup.

Hot Cereals

ANY OF THESE RECIPES can be dressed up with nuts, brown or white sugar, milk, or dried fruit from the pantry. Be creative!

Hot Barley Cereal, page 106

Apricot Honey Oatmeal

3½ cups water

RESEALABLE BAG:
  ½ cup chopped dried apricots
  ½ tsp. ground cinnamon

PANTRY ITEM:
  ⅓ cup honey

RESEALABLE BAG:
  2 cups oats

IN MEDIUM SAUCEPAN OVER high heat, bring water, apricots, honey, and ground cinnamon to a boil. Reduce heat to medium and stir in oats; return to a boil. Cook for about 5 minutes, stirring frequently. To serve, add nuts or dried fruit and milk to taste.

Six-Grain Cereal

6 cups water

RESEALABLE BAG:
  4 cups 6-grain cereal

IN A MEDIUM SAUCEPAN, over high heat, bring water to a rolling boil. Reduce heat to medium and add cereal. Stirring occasionally, cook cereal until soft, about 20 to 25 minutes. To serve, add sugar, nuts, or dried fruit and milk to taste.

Quinoa

4 cups water

RESEALABLE BAG:
  3 cups quinoa

IN A HEAVY SAUCEPAN* over medium heat, combine water and quinoa. Stirring occasionally bring water and quinoa to a boil. Reduce heat to medium-low and simmer for 20 minutes or until quinoa is soft. To serve, add sugar, nuts or dried fruit and milk to taste.

*MY PREFERRED METHOD FOR cooking quinoa is in a rice cooker.

Brown Rice Breakfast

(Original recipe from the kitchen of Anne Stanger)

3 cups cooked brown rice
2 (12-oz.) cans evaporated milk

RESEALABLE BAG:
  1¼ tsp. ground cinnamon
  5 Tbsp. raisins
  4 tsp. chopped nuts

PANTRY ITEMS:
  ¾ cup maple syrup

PLACE ALL INGREDIENTS INTO a small saucepan and warm over medium heat.

Hot Millet Cereal

4 cups water

RESEALABLE BAG:
  1 cup cracked millet
  12 Tbsp. powdered milk

PLACE MILLET IN BLENDER and pulse until millet is the consistency of uncooked farina*. In a medium saucepan combine powdered milk and water and bring to a boil over medium heat. Once water and milk are boiling, stir continuously while adding cracked millet. When millet cereal is thick, about 5 minutes, add fruits, nuts, milk, and brown sugar to taste.

*FARINA, SEMOLINA, GERMADE, AND Cream of Wheat® are essentially the same thing: finely ground wheat cereals that are served hot.

Hot Barley Cereal

6 cups water

RESEALABLE BAG:
  1½ cups barley

IN HEAVY SAUCEPAN WITH a tight fitting lid, boil water and barley for 45 minutes over a medium-low heat. To serve, add sugar, nuts or dried fruit and milk to taste.

Oatmeal

6 cups water

RESEALABLE BAG:
3 cups oats
¾ tsp. salt

IN A MEDIUM SAUCEPAN, bring water and salt to a rolling boil on high heat. Reduce heat to medium. Add oatmeal. Cook 5 minutes. To serve, add brown sugar, nuts, or dried fruit and milk to taste.

Farina*

4 cups water

RESEALABLE BAG:
1 cup farina*
¾ cup powdered milk
½ tsp. salt

PANTRY:
milk
brown sugar

IN A MEDIUM SAUCEPAN, bring water and salt to a boil over high heat. Reduce heat to medium and stir in farina and powdered milk. Boil for 3 minutes, stirring continuously. Serve with milk and brown sugar to taste.

*FARINA, SEMOLINA, GERMADE, AND Cream of Wheat® are finely ground wheat cereals that are served hot.

Cracked Wheat

3 cups water

RESEALABLE BAG:
1 cup cracked wheat

PANTRY:
milk
brown sugar

COMBINE ALL INGREDIENTS IN a heavy saucepan. Cook over medium heat, stirring frequently, until wheat is soft, about 30 minutes. Serve with milk and brown sugar to taste.

Breakfast Couscous

2 cups water

RESEALABLE BAG:
2 cups couscous
6 Tbsp. powdered milk
3 tsp. ground cinnamon

RESEALABLE BAG:
½ cup sliced almonds
⅓ cup dried apricots
⅓ cup raisins

PANTRY ITEM:
2 Tbsp. honey
milk

IN A MEDIUM SAUCEPAN over medium-high heat, combine water, powdered milk, honey, and ground cinnamon. When water boils, stir in couscous. Remove from heat and let stand, while covered, for 15 minutes. Add apricots, raisins, and almonds. Serve with milk to taste.

Rice Pudding

3½ cups water, divided

RESEALABLE BAG:
  1 cup rice

RESEALABLE BAG:
  7 Tbsp. powdered milk
  ½ cup granulated white sugar
  ¼ cup raisins
  1 tsp. ground cinnamon
  1 tsp. orange zest
  3 Tbsp. sweetened coconut

TO COOK RICE: IN a heavy bottomed saucepan with a tight fitting lid, place 1 cup water and add 1 cup rice. Boil on high heat for 15 minutes. Remove from heat and set aside for 10 minutes, then fluff. Do not lift the lid while rice is resting. If you have a rice cooker, just add water and rice and depress the lever. Fluff rice when lever pops up. Combine all ingredients in a medium saucepan. Stirring frequently, cook for 30 minutes over low heat. Rice pudding can be served warm or cold.

Muffins

A WORD ABOUT MUFFINS: If you want light, fluffy muffins, create a well with the dry ingredients and add liquids in the center. Stir only until moist. Overmixing creates tough, elastic muffins.

ONE MORE THING: If your recipe calls for brown sugar, make sure you combine it with other ingredients when you bag your mix or you will have a hard lump of brown sugar as time passes.

Chocolate Chip Muffins, page 119

Pumpkin Pie Muffins

1 cup canned pumpkin puree
½ cup + 1 Tbsp. water

RESEALABLE BAG:
1 cup flour
½ cup flax seed meal (If you don't have flax seed meal, substitute flour)
1 Tbsp. baking powder
1¼ tsp. pumpkin pie spice
½ tsp. baking soda
½ tsp. salt
⅓ cup powdered milk
1 Tbsp. powdered egg
1 cup raisins

PANTRY ITEMS:
⅓ cup oil
½ cup honey
vegetable shortening

PREHEAT OVEN TO 400°F. Grease muffin pan cups with vegetable shortening. In a large bowl, mix all ingredients until moist. Fill muffin cups ⅔ full of batter. Bake 15 to 20 minutes or until a toothpick inserted into the center of a muffin comes out clean. Cool muffins 5 minutes before removing from cups.

Bran Muffins

¾ cup water
3 Tbsp. cooled melted canned butter

RESEALABLE BAG:

1 cup flour
3 Tbsp. granulated white sugar
1 Tbsp. powdered egg
3½ tsp. baking powder
¼ tsp. salt
3 Tbsp. powdered milk
1 cup bran

PANTRY ITEM:

vegetable shortening

PREHEAT OVEN TO 375°F. Grease muffin pan cups with vegetable shortening. In a medium bowl, mix all ingredients until moist. Fill muffin cups ⅔ full of batter. Bake for 20 to 25 minutes or until a toothpick inserted into the center of a muffin comes out clean. Cool muffins 5 minutes before removing from cups.

Spiced Muffins

1¼ cup water
⅓ cup oil

RESEALABLE BAG:
  2 cups flour
  ½ tsp. salt
  2 tsp. ground cinnamon
  1 Tbsp. powdered eggs
  3 Tbsp. powdered milk
  2 tsp. baking powder
  ½ cup brown sugar
  ¼ cup granulated white sugar
  ½ tsp. ground cloves
  ¼ tsp. ground ginger
  ¼ tsp. allspice
  ¾ cup chunky cinnamon applesauce

PANTRY ITEM:
  vegetable shortening

PREHEAT OVEN TO 375°F. Grease muffin pan cups with vegetable shortening. In a medium bowl, mix all ingredients until moist. Fill muffin cups ⅔ full of batter. Bake for 20 to 25 minutes or until a toothpick inserted into the center of a muffin comes out clean. Cool muffins 5 minutes before removing from cups.

Blueberry Muffins

¾ cup + 2 Tbsp. water
½ cup melted canned butter or safflower oil

RESEALABLE BAG:
2 cups flour
¾ cup granulated white sugar
2 tsp. baking powder
½ tsp. salt
1 Tbsp. powdered egg
2 Tbsp. powdered milk

PANTRY ITEM:
vegetable shortening
1 tsp. vanilla extract

RESEALABLE BAG:
1 cup dried blueberries (For soft blueberries,
   cover the night before with 1 cup water. Drain
   in the morning and gently fold into batter.)

PREHEAT OVEN TO 375°F. Grease muffin pan cups with vegetable shortening. In a large bowl, mix all ingredients until moist, except blueberries. Fold in blueberries. Fill muffin cups ⅔ full of batter. Bake 20 to 25 minutes or until a toothpick inserted in the center of a muffin comes out clean. Cool muffins 5 minutes before removing from cups.

Cheddar Cheese & Sausage Muffins with Gravy

1 pint sausage, crumbled
1 (10.75-oz.) can condensed cheddar cheese
   soup, do not reconstitute.
1 cup water
¼ cup melted canned butter

RESEALABLE BAG:
   3 Tbsp. powdered milk
   1 tsp. ground sage
   2 cups all-purpose flour
   2 tsp. baking powder

PANTRY ITEM:
   vegetable shortening

PREHEAT OVEN TO 350°F. Grease muffin pan cups with vegetable shortening. In a large bowl mix all of the ingredients, except sausage, until moist. Add crumbled sausage. Spoon muffin batter into prepared muffin cups. Bake 25 minutes. Cool 5 minutes before removing muffins from cups. While muffins are baking, prepare gravy.

GRAVY:
   1½ cups liquid (For a richer gravy, use sausage drippings and water to equal 1½ cups total liquid. For a reduced fat version, just use water.)

RESEALABLE BAG:
   5 Tbsp. powdered milk
   2 Tbsp. cornstarch
   1 tsp. salt
   ⅛ tsp. black pepper, or to taste

IN A MEDIUM SAUCEPAN, stir flour, powdered milk, and ½ cup liquid together until smooth. Add salt, pepper, and remaining water. Cook, stirring frequently, on medium heat until gravy is thick and bubbly. Serve gravy over muffins.

Multi-grain Muffins

¾ cup + 1 Tbsp. evaporated milk

RESEALABLE BAG:

½ cup all-purpose flour

½ cup wheat flour

½ cup rye flour

½ cup cornmeal

¾ cup oats

¼ cup granulated white sugar

1 tsp. baking soda

½ tsp. salt

1 Tbsp. powdered egg

PANTRY:

⅓ cup oil

⅓ cup honey

2 Tbsp. molasses

vegetable shortening

RESEALABLE BAG:

1 cup raisins

PREHEAT OVEN TO 375°F. Grease muffin pan cups with vegetable shortening. In a large bowl mix all ingredients, except raisins, until moist. Mix in raisins. Fill muffin cups ⅔ full of batter. Bake 20 to 25 minutes or until a toothpick inserted in the center of a muffin comes out clean. Cool muffins 5 minutes before removing from cups.

Gingerbread Muffins

1 cup + 2 Tbsp. evaporated milk

RESEALABLE BAG:

2 cups all-purpose flour

½ cup whole wheat flour

½ cup brown sugar

2 tsp. ground ginger

1 tsp. salt

1 tsp. ground cinnamon

⅛ tsp. ground cloves

⅛ tsp. ground nutmeg

2 Tbsp. powdered egg

PANTRY ITEMS:

½ cup oil

¼ cup molasses

vegetable shortening

RESEALABLE BAG:

1 cup currants or raisins

PREHEAT OVEN TO **375°F.** Grease muffin pan cups with vegetable shortening. In a large bowl, stir together all ingredients, except currants or raisins. When batter is smooth, add currants or raisins. Fill muffin cups ⅔ full of batter. Bake for 20 to 25 minutes or until toothpick inserted into the center of muffin comes out clean. Cool muffins 5 minutes before removing from cups.

Chocolate Chip Muffins

RESEALABLE BAG:
2 cups flour
1/3 cup firmly packed brown sugar
1/3 cup granulated white sugar
2½ tsp. baking powder
¾ tsp. salt
5 Tbsp. powder milk
2 Tbsp. powdered eggs

1¼ cups water
½ cup cooled melted butter

RESEALABLE BAG:
1 cup milk chocolate chips
½ cup coarsely chopped almonds

PANTRY ITEMS:
2 tsp. vanilla extract
vegetable shortening

PREHEAT OVEN 375°F. GREASE muffin pan cups with vegetable shortening. Pour dry ingredients from first resealable bag into a large bowl. Make a well in the center of dry ingredient mixture. Stir in water, melted butter, and vanilla until moist. Fold in chocolate chips and nuts. Spoon batter into muffin cups. Bake 20 minutes or until a tooth pick inserted into the middle of a muffin comes out clean. Cool muffins 5 minutes before removing from cups.

Tropical Muffins

1 (13.5-oz.) can coconut milk
½ cup liquid (Use the juice from the pine-
    apple and add additional water if needed.)
1 (4-oz.) can crushed pineapple, reserve juice

RESEALABLE BAG:
2½ cups flour
1 cup granulated white sugar
½ tsp. salt
4 tsp. baking powder

RESEALABLE BAG:
1 cup sweetened coconut flakes

PANTRY ITEM:
1 Tbsp. coconut extract
vegetable shortening

PREHEAT OVEN TO 375°F. Grease muffin pan cups with vegetable shortening. In a large bowl, mix all ingredients, except coconut and crushed pineapple, until moist. Fold in coconut and crushed pineapple. Fill muffin cups ⅔ full of batter. Bake 20 to 25 minutes or until a toothpick inserted in the center of a muffin comes out clean. Cool muffins 5 minutes before removing from cups.

Crumb Cake Muffins

MUFFIN TOPPING:

⅓ cup softened canned butter

RESEALABLE BAG:
¾ cup flour
½ cup brown sugar
½ tsp. cinnamon

BATTER:

RESEALABLE BAG:
1¾ cups flour
¾ cup granulated white sugar
1¾ tsp. baking powder
½ tsp. salt
2 Tbsp. powdered egg
3 Tbsp. powdered milk

1 cup + 2 Tbsp. water
½ cup cooled melted canned butter

PANTRY ITEMS:
2 tsp. vanilla extract
½ tsp. almond extract
vegetable shortening

PREHEAT OVEN TO 375°F. Grease muffin pan cups with vegetable shortening. In a small bowl, combine topping ingredients. Stir topping ingredients until they resemble coarse crumbs. Pour dry ingredients into a large bowl and form a well. Add water, butter, vanilla, and almond extract. Stir just enough to combine dry ingredients with the wet ingredients. Do not over mix. Fill muffin cups ⅔ full of batter. Top muffin batter with prepared topping. Lightly spritz muffin topping with water. This will help adhere crumb topping to muffin batter. Bake for 20 to 25 minutes or until a toothpick inserted into the middle of a muffin comes out clean. Cool muffins 5 minutes before removing from cups.

Wheat-germ Cornbread Muffins

1¾ cups water
⅓ cup melted canned butter

RESEALABLE BAG:
1 cup whole wheat flour
1 tsp. salt
⅓ cup granulated white sugar
5 tsp. baking powder
1 cup wheat-germ
1 cup cornmeal
3 Tbsp. powdered eggs

PANTRY ITEM:
vegetable shortening

PREHEAT OVEN TO **400°F.** Grease muffin pan cups with vegetable shortening. In a large bowl mix all ingredients, except raisins, until moist. Fold in raisins. Fill muffin cups ⅔ full of batter. Bake 20 to 25 minutes or until a toothpick inserted in the center of a muffin comes out clean. Cool muffins 5 minutes before removing from cups.

Miscellaneous Breakfasts

Homemade Granola

(Original recipe from the kitchen of Belinda Craft)

RESEALABLE BAG:
1 cup oat bran or wheat bran
1 cup coconut
½ cup sesame seeds
1 cup wheat germ
1 cup wheat flour
½ cup sunflower seeds
¾ cup chopped nuts
1 tsp. salt

PANTRY ITEMS:
1 cup oil
1 cup honey
1 cup peanut butter
1 Tbsp. vanilla extract

PREHEAT OVEN TO 200°F. In a large bowl combine ingredients. Spread on baking sheet and bake for 2 hours.

Breakfast Bars

PANTRY ITEMS:
- 1 cup honey
- 4 Tbsp. canned butter
- 2/3 cup peanut butter
- 1 Tbsp. vanilla extract

RESEALABLE BAG:
- 5 cups oat meal
- 2 Tbsp. sesame seeds
- 2 Tbsp. sliced almonds
- 2 Tbsp. pumpkin seeds
- 2 Tbsp. sunflower seeds
- 1/4 cup wheat germ
- 1/4 cup shredded coconut
- 1/2 cup white chocolate chips
- 1/3 cup dried cherries
- 1/3 cup dried blueberries
- 1/3 cup dried cranberries

IN A MEDIUM PAN, boil honey and butter for 1 minute. Remove from heat and add peanut butter and vanilla extract. Stir until smooth. Mix remaining ingredients with peanut butter mixture. On a cookie sheet, press dough into a rectangle one-inch high. Cool and cut into bars.

Savory Pancakes with Sausage and Cheese

2 cups warm water

RESEALABLE BAG:
2 chicken bouillon cubes
2 Tbsp. dried celery
3 Tbsp. dried bell peppers
1 Tbsp. dried onion
½ Tbsp. dried dill
¼ tsp. turmeric
1 tsp. granulated white sugar
3 Tbsp. powdered milk

RESEALABLE BAG:
2 cups flour

PANTRY ITEM:
vegetable oil

1 fried egg, if available
1 pint canned sausage patties
1 (8-oz.) canned cheddar cheese, grated

IN A LARGE BOWL, combine all ingredients, except flour, and allow vegetables to rest for 10 minutes, or until soft. Stir in flour. Ladle pancake batter onto a hot oiled griddle. Turn pancakes when edges are brown and bubbles appear. To serve, place hot pancake on plate and top with a warm sausage patty, fried egg, and grated cheese.

Orange Marmalade Rolls

1 cup + 3 Tbsp. warm water
¼ cup warm water
6 Tbsp. canned butter, melted
1 pint orange marmalade

RESEALABLE BAG:
¼ tsp. granulated white sugar

RESEALABLE BAG:
2 cups whole wheat flour

RESEALABLE BAG:
3 Tbsp. powdered milk
3 Tbsp. powdered egg
2 tsp. salt
2 tsp. lemon zest
½ tsp. ground ginger

RESEALABLE BAG:
2½ cups all-purpose flour

PANTRY ITEM:
1½ Tbsp. yeast
1⅓ cup maple syrup

PREHEAT OVEN TO 350°F. In a small bowl, combine ¼ cup water, yeast, and sugar. In a large bowl, combine wheat flour, water, butter, maple syrup, powdered milk, powdered eggs, salt, lemon zest, and ginger. Stir until smooth. Add yeast mixture and gradually add in all-purpose flour until a soft—but not sticky—dough forms. You may not need all of the flour (I add about ¼ cup at a time). Knead bread on a floured surface until dough is smooth and elastic, about 5 minutes.

COAT THE SIDES AND bottom of a large bowl with a tablespoon of oil. Roll dough in bowl to lightly coat in oil. Cover with plastic wrap and let rise in a warm area until double in size, usually about 1 hour. Roll out risen dough into a rectangle ¼-inch thick. Spread orange marmalade over dough, leaving a 1-inch border. Roll dough like a jelly roll. Pinch ends to seal. Slice in 2-inch segments. Transfer rolls to greased 11" x 13" baking dish. Bake rolls for 30 minutes or until golden brown. Cool before serving.

Hot Ham Rolls

RESEALABLE BAG:
2 cups all-purpose flour
3 tsp. baking powder
½ tsp. salt
3 Tbsp. powdered milk

PANTRY ITEM:
4 Tbsp. shortening

1 cup water

2 cups canned ham
1 (10.75-oz.) can condensed mushroom soup
¼–½ cup water

RESEALABLE BAG:
1 Tbsp. dried parsley

PREHEAT OVEN TO 375°F. In a medium bowl, combine flour, baking powder, salt, shortening, powdered milk, and 1 cup water. Roll dough out into a rectangle ⅓ inch thick. Mix ham with enough mushroom soup to form a paste. Spread paste on dough. Roll up jelly roll–style. Cut into 1-inch pieces and lay cut side up in a greased 9" x 13" baking dish. Bake 25 minutes. In a small saucepan over medium heat, mix remaining mushroom soup with enough water to form gravy. To serve, place roll on plate, spoon mushroom soup over roll and sprinkle with parsley.

Best-Ever Bread Pudding

8 slices dried bread
½ cup raisins
1 (12-oz.) can evaporated milk
2 cups ricotta cheese
½ cup sugar
2 eggs
1 cup yogurt
1 tsp. lemon zest
1 tsp. vanilla
¼ tsp. cinnamon

CUBE BREAD AND TOSS with raisins. Place bread mixture in soufflé dish. Mix remaining ingredients and pour over bread and raisins. Place soufflé dish in slow cooker with 2 inches boiling water. Cook on high for 2.5 hours.

SECTION TWO:
COOL STUFF YOUR MAMA NEVER TAUGHT YOU

Do-It-Yourself
Dairy Recipes

Butter

1 quart cream
salt to taste

PLACE CREAM IN FOOD processor and select highest speed. Watch the cream. It will first become the texture of whipping cream. Soon after, globules of butter fats will form within the cream. Once the butter has separated from the liquid, stop processing. In my processor this process only takes about 5 minutes. What you will see is butter in "light milk." The "light milk" is buttermilk. I use this for cooking, drinking, or in my compost pile.

WASH YOUR HANDS IN cold water. Take a handful of butter out of the processor and gently squeeze the butter between your hands to remove remaining buttermilk. It only takes 2 or 3 squeezes. Place butter in a bowl and salt to taste. Place in sealed containers and refrigerate or freeze. Makes 2 cups.

Buttermilk

4 cups milk*
1 cup cultured store-bought buttermilk

IN A MICROWAVEABLE DISH, warm milk to 90°F. (I heat each cup for 50 seconds in the microwave.) Pour warm milk into a container and stir in buttermilk. Leave on counter for 12 hours. Makes 1¼ quarts.

REMEMBER TO RESERVE 1 cup buttermilk as your starter for your next batch; if you always reserve one cup, you will never need to buy buttermilk again. A starter will keep in the refrigerator for 1–3 weeks and up to 3 months if frozen.

*Do not use ultra-pasteurized milk.

Yogurt

(Original Recipe from the Kitchen of Tammy Hulse)

1 gallon 2% or whole milk*
5 Tbsp. powdered milk
1 cup plain yogurt
sweetener of choice**

PREHEAT OVEN ON WARM setting. Place milk and powdered milk in a large thick-bottomed stainless steel pot and heat milk to 180°F, stirring often. Turn off oven and turn on oven light. Once the milk has reached 180°F, cool to 125–110°F. (I place the pot in a sink full of ice water; within 5 to 10 minutes, the milk cools between 125–110°F.) Stir in yogurt and sweetener. Pour into pint-sized canning jars and cover with a canning lid and ring or aluminum foil. Remember to save 1 cup homemade yogurt as a starter for your next batch.

PLACE FILLED JARS IN oven and allow to ripen to desired consistency. I ripen mine 4 to 6 hours. Place yogurt in refrigerator. Makes 9 pints.

YOGURT WILL KEEP IN refrigerator for a good four weeks.

*Do not use ultra-pasteurized milk or yogurt will not set.

**I sweeten my yogurt by adding ¾ cup organic fructose to 1 gallon of milk. You may choose to add more or less sweetener depending on your preference. I add the fructose with the powdered milk.

Sour Cream

1 cup cream
2 Tbsp. buttermilk.

HEAT CREAM TO 90°F (I heat cream in the microwave for 50 seconds). Pour warm cream into a container and stir in buttermilk. Pour in canning jar and cover with a canning lid. Set on counter for 12 to 24 hours until very thick.

IN THE WINTER, I place my sour cream mixture in the oven and turn on the oven light for added warmth.

MAKES 8 OUNCES.

Yogurt Cheese

32-oz. yogurt at room temperature
½ tsp. non-iodized salt such as sea salt or flake
    cheese salt

LINE A COLANDER WITH cheese cloth. Place colander over a large container to catch the whey, and spoon yogurt into colander. Periodically, twist and squeeze the cheesecloth to accelerate the draining of the whey. Once yogurt cheese has reached desired consistency (I usually drain mine overnight), fold in salt. Makes 8 ounces.

YOGURT CHEESE CAN BE stored in an airtight container in the refrigerator for 2 to 3 weeks. This is delicious on bagels. Add dried strawberries or any other dried fruit for a special bagel spread.

Lemon Cheese Spread

½ gallon whole milk*
¾ tsp. non-iodized salt, such as sea salt or flake
   cheese salt
¼ cup lemon juice**

PLACE MILK IN A large, thick-bottomed stainless steel pot. Heat the milk slowly to 165°F, stirring frequently to avoid scorching. Once the temperature is reached, remove pot from heat. Add lemon juice and let sit undisturbed for 15 minutes. Line colander with cheesecloth. Place a large container under the colander to catch whey. Pour cheese curds and whey into colander. Allow curds to drain until center of curds is no longer hot to the touch. Twist and squeeze cheesecloth to accelerate draining of the whey. With a spoon, mix in salt. Makes about 1 pound.

THIS CHEESE CAN BE stored in an airtight container for 2 to 3 weeks. For a change of pace, we add herbs, nuts, canned salmon, or dried fruit and serve with crackers.

*Recipe will not work with ultra-pasteurized milk
**When I haven't had lemon juice, I have used vinegar. The results were acceptable.

Paneer

10 cups milk
¼ cup lemon juice.

BRING MILK TO A boil in a large, thick-bottomed stainless steel pot. Once boiling, remove milk from heat and add lemon juice. Stir gently and set aside, until curds separate from whey. Drain in a colander lined with cheesecloth. Makes about one pound.

BECAUSE MANY PEOPLE ARE unfamiliar with paneer cheese and how to use it, I have included below a favorite recipe using paneer cheese.

Saag Paneer

1 cup fresh or frozen spinach
8 Tbsp. melted butter
1 tsp. powdered cumin
1 tsp. garlic powder
2 tsp. ground coriander
1 tsp. red chili powder
3 Tbsp. vegetable oil
1 batch paneer cheese
½ cup cream
Salt to taste

COOK SPINACH IN BUTTER. Add spices and sauté in oil. Add paneer and stir in cream. Serve immediately with flat bread or warmed tortillas.

Ricotta

1 gallon whole milk*
½ tsp. sea salt or flake cheese salt
2 tsp. citric acid powder**
½ cup cool water

IN A SMALL BOWL, combine citric acid and water. Place milk in a large, thick-bottomed stainless steel pot. Place pot filled with milk into another larger pot, which is partially filled with water. It will look like a large double boiler. Slowly heat milk to a target temperature of 190°F. Stir frequently to avoid scorching. Once temperature has been reached, remove milk from heat. Stir in half of the citric acid solution. Let milk sit undisturbed for 3 minutes.

LINE A COLANDER WITH cheesecloth. Place a large container under the colander to catch whey. Scoop out curds with a slotted spoon. Place curds in colander. When all curds are scooped out, return milk to burner and heat to 180°F. Stir in remaining citric acid solution. After 2 to 3 minutes of gentle stirring, scoop out remaining curds and place in colander. The whey should now be clear yellow, not milky white. Allow cheese curds in colander to drain until cheese is warm to the touch. Twist and squeeze the cheese cloth to accelerate the draining of the whey.

WHEN WHEY NO LONGER drips from the curds, place cheese in a bowl and, with a fork, blend in salt, breaking up any large curds in the process. Ricotta can now be used or stored in an airtight container in the refrigerator for 2 to 3 weeks. Makes about one pound

*Recipe will not work with ultra-pasteurized milk.
**When I haven't had citric acid, I have used vinegar. The results were acceptable.

Combination Mozzarella and Ricotta

Phase 1: Mozzarella

1 gallon whole milk*
1½ tsp. citric acid powder**
¼ cup cool water
1 tsp. non-iodized salt, such as sea salt or flake
    cheese salt

RENNET SOLUTION:

½ rennet tablet***
¼ cup cool water
4 cups cold brine solution****

IN A SMALL BOWL, combine citric acid and ¼ cup cool water. In a separate bowl, dissolve half rennet tablet into remaining ¼ cup cool water. Heat milk in a large, thick-bottomed stainless steel pot to 60°F. While stirring slowly, add citric acid solution. Continue to heat milk up to 90°F. Remove from heat and stir in rennet solution.

ALLOW MILK TO SIT undisturbed for at least 40 minutes. The whey should be clear and yellow, not white and milky. Line a colander with cheesecloth. Place a large (greater than one gallon) container under colander to catch the whey. Pour curds and whey into colander, reserving whey. Tie up cheese cloth and place a wooden spoon through the knot.

HANG CHEESECLOTH BAG OVER whey dripping bowl for about 10 minutes or until warm to the touch. Remove bag from wooden spoon and twist to remove additional whey. In a small bowl, mix salt into cheese.

MICROWAVE CHEESE ON HIGH for 1 minute, then knead hot cheese with a spoon. Repeat until cheese has a warm taffy feel and a glossy sheen. Roll cheese into ball. Immediately place cheese in cold brine solution until ricotta is in Phase 2.

ONCE RICOTTA HAS BEEN MADE, rinse mozzarella under cool water to remove excess brine. Store in an airtight container in the refrigerator for 2 to 3 weeks. Makes one pound.

Phase 2: Ricotta

½ gallon whole milk*

½ tsp. non-iodized salt, such as sea salt or flake cheese salt

1 ½ tsp. citric acid

½ cup cool water

POUR LEFTOVER WHEY FROM mozzarella back into original pot. Add ½ gallon milk to reserved whey. In a small bowl, combine citric acid and water. Place milk in a large, thick-bottomed stainless steel pot.

PLACE POT FILLED WITH milk into another large pot filled with water. It will look like a giant double boiler. Slowly heat milk to 190°F, stirring frequently to avoid scorching, and then remove milk from heat. Stir in half of citric acid solution. Let milk sit undisturbed for 3 minutes. Line a colander with cheesecloth. Place a large container under the colander to catch whey. After milk has rested for 3 minutes, scoop out curds with a slotted spoon. Place curds in colander and return milk to burner. Heat milk to 180°F, and stir in remaining citric acid solution. After 2 to 3 minutes of gentle stirring, scoop remaining curds into colander. Note: whey should now be clear yellow, not milky white.

ALLOW CHEESE CURDS IN colander to drain until cheese is warm to the touch. Twist and squeeze cheesecloth to accelerate draining of the whey. When whey no longer drips from curds, transfer cheese to a bowl and mix in salt with a fork. Store ricotta in an airtight container in the refrigerator for 2 to 3 weeks. Makes about one and a half pounds.

*Recipe will not work with ultra pasteurized milk.

**Substitute vinegar for citric acid, if necessary.

***Rennet is found in most grocery stores on the pudding and custard isle.

****Brine solution is made by dissolving ⅓ cup salt into 4 cups water.

I Can; You Can; We All Can: Home Canning Basics

Terms to Know

Agricultural Extension:

An agricultural extension is a federally operated agricultural network that provides scientific, educational, and consumer resources at little or no charge. As a consumer, you will have access to expert knowledge such as gardening, food preservation, finance, and emergency preparedness. Many agricultural extensions will even test pressure canner gauges free of charge or for a nominal fee. To find the nearest agricultural extension in your state visit, http://www.csrees.usda.gov/Extension/ and remember the old adage, an ounce of prevention is worth a pound of cure.

Boiling Water Bath Time Table:

To effectively process foods in a boiling water bath, foods must be processed for a specific number of minutes under boiling water that is at least 2 inches above the jar lid. The number of minutes that food must be processed is determined by the altitude where the food will be processed. If your friend lives in California at sea level, and a boiling water bath processes a food item in 20 minutes, and you are living in Colorado, you will discover that you must process for a longer period of time.

Botulism:

Botulism is caused by a bacterium named Clostridium botulinum. It is commonly found in soil. According to the Centers for Disease Control and Prevention, an average of 145 cases of botulism are reported each year in the Unites States, with 15 percent of the cases caused by foods that were incorrectly processed by home canners.

Botulism can be treated; however, severe cases have resulted in death. The signs and symptoms for botulism may include double and/or blurred vision, drooping eyelids, slurred speech, difficulty swallowing, dry mouth, and muscle weakness. If you should experience any of these symptoms, be sure to tell the emergency room physician that you are a home canner, and, if possible, provide the hospital with the jar and any leftover food. Through testing the jar and food, the hospital can determine if Clostridium botulinum bacteria are present.

For more information on botulism, please visit, http://www.cdc.gov/nczved/dfbmd/disease_listing/botulism_gi.html.

Cold Pack:

The food item and liquid or syrup is cold when placed in the canning jar.

Head Space:

The amount of space below the top rim of the canning jar.

Hot Pack:

The food item and liquid, syrup, or broth is heated to boiling, prior to being placed in the canning jar.

Time and Pressure Table:

To effectively process foods in a pressure canner, one must use a time and pressure table to determine the number of minutes that food needs to be processed at a specific number of pounds of pressure. A time and pressure table will be included in the pressure canner instruction booklet that came with your pressure canner.

Times and pressures are determined by the altitude where the food item will be pressure cooked. If your friend lives in California and pressure cooks something for 10 minutes at 10 pounds of pressure, and you are living in Colorado, you will discover that the time and pressure tables will have you pressure cook the same item at a higher poundage.

If you are water-bath processing, the same principle holds; you must increase the processing time when altitude increases.

Vegetables:

Select fresh, firm vegetables. Do not bottle overripe vegetables because they will have a soft texture and will be at a greater risk for spoilage. Vegetables must be processed in a pressure canner. It is important that you carefully read the recipe and process accordingly, as the risk of botulism increases with vegetables not correctly acidified or processed. When using a pressure canner, thoroughly read the included instructions for safe operation and to determine how long and how many pounds of pressure should be used. The head space for most vegetables is ½-inch; however, corn, peas, and lima beans require a 1-inch head space. Remember, vegetables, including tomatoes, must be boiled for 10 minutes before eating!

Fruits:

Select fresh, firm fruits. Do not bottle overripe or bruised fruits because they will have a soft texture and will be at a greater risk for spoilage. Fruits should be processed in a boiling water bath. Fruits may be cold packed or hot packed. The head space for fruits is ½ to 1½ inches, depending on the fruit.

Meats, Poultry, and Fish:

Meat, poultry, and fish should never be coated in flour, bread crumbs, and so forth before being bottled, because this adds to the risk of food poisoning. Meat, poultry, and fish must be processed in a pressure canner. When using a pressure canner, thoroughly read the included instructions for safe operation and to determine how long and how many pounds of pressure should be used for processing. You may raw pack or hot pack according to the recipe.

Raw-packed meats, poultry, or fish should not have any liquid added to the jar prior to processing. If the meat, poultry, or fish has been precooked prior to bottling, you may add 3 to 4 tablespoons of water, juice, or broth. The head space for meat, poultry, or fish is 1 inch. Remember, home-canned meats, poultry, and fish must be boiled for 10 minutes before eating!

Basic Instructions

1. Examine canning jars for nicks or cracks and discard if any are found.
2. Wash jars in hot, soapy water or in dishwasher.
3. Place canning lids in a pan of hot water. Keep water scalding hot until ready to use, but do not boil.
4. Prepare food according to recipe.
5. Pack food into jar, leaving a ½-inch head space for vegetables. Leave a 1-inch head space for corn, peas, lima beans, meats, poultry, and fish. Fruits should have a ½-to 1½-inch head space, depending on type of fruit being canned.
6. Add liquid.
7. Fruits—add boiling syrup over fruit, leaving a ½- to 1½-inch head space, depending on type of fruit being canned.
8. Vegetables—add boiling water and salt, leaving a ½-inch head space.
9. Meat, poultry, and fish—For precooked meat, poultry, or fish, add 3 or 4 tablespoons of liquid or broth. When raw processing, do not add liquid. For both raw and precooked meat, poultry, and fish, leave a 1-inch head space.
10. Using a clean cloth, wipe rim of jar clean of syrup, seeds, juice, broth, and so on.
11. Place canning lids on top of filled jars. Rubberized ring goes next to jar lip.
12. Place screw band on top of lid and jar. Firmly tighten by hand.

13. Process according to recipe, altitude, and timetable instructions included with your canner Be sure jars do not touch canning wall while processing, or jars may break.

14. Remove jars from canner and place 2 to 3 inches apart on a cooling rack or several layers of towels to cool. Do not set jars in cool drafts or on a cold or wet surface, because bottles may break.

15. When jars have cooled, remove screw bands. Test to make sure they have properly sealed by pressing on the center of the lid. If the center of the lid does not move when depressed, the jar is sealed. If there is any movement, the jar did not seal and contents should be refrigerated and eaten within a short period of time.

16. Mark on top of lid with a permanent marker the date that the food was processed and what the food is. For example, "1/2010—Roast beef."

Vegetable Recipes

Corn

(May cold pack)

corn
water
1 tsp. salt per pint jar
1 tsp. sugar per pint jar (optional)

USE ONLY CORN THAT is at its peak of freshness. Shuck corn and remove corn silk. Wash corn. Cut corn from cob. Pack corn into prepared canning jars with a 1-inch head space. Pour salt, sugar, and boiling water into jar, leaving a ½-inch head space. Wipe jar lip clean. Place lid on jar. Place screw band on jar and firmly tighten by hand. Corn must be pressure canned according to the instruction booklet's time and pressure table for your altitude.

Green Beans

(May cold pack)

green beans
1 tsp. salt per quart jar
water

WASH, STRING, AND CUT green beans to desired length. Pack into prepared canning jar, leaving a ½-inch head space and pour salt and boiling water into jar. Wipe jar lip clean and place on jar. Place screw band on jar and firmly tighten by hand. Green beans must be pressure canned according to the instruction booklet's time and pressure table for your altitude.

Stewed Tomatoes

(May cold pack. And, I don't care what anyone says, tomatoes are a vegetable in my book!)

    2 quarts tomatoes*
    4 Tbsp. chopped green peppers
    4 Tbsp. chopped onions
    6 Tbsp. chopped celery
    2 tsp. celery salt
    2 tsp. sugar
    ¼ tsp. salt

WASH TOMATOES. PLACE TOMATOES in a scalding hot pan of water until skins begin to split. Immediately plunge tomatoes into cold water. Slip tomato skins off of tomato. Cut tomato into chunks and place into a large bowl. Combine remaining ingredients and gently stir. Pack into prepared canning jars, leaving a 1-inch head space. Wipe jar lip clean and place lid on jar. Place screw band on jar and firmly tighten by hand. Stewed tomatoes can be boiling-water-bath processed according to the instruction booklet's timetable for your altitude.

*IF THE TOMATOES YOU are using for canning are of the low acid varieties, add 1 tablespoon lemon juice or vinegar to each quart. The lemon juice or vinegar will increase the acidity level, which will reduce the risk of botulism.

Tomato Soup

7 quarts skinned tomatoes
3 medium onions, chopped
3 stalks celery, chopped
14 sprigs fresh parsley
½ cup sugar
3 Tbsp. salt
2 tsp. black pepper

TO SKIN TOMATOES, PLACE tomatoes in hot water until skins split. Immediately place in cold water. Skins will slip off. Place all ingredients in a large stock pot over medium-high heat, stirring frequently. Cook until all vegetables are tender. Carefully spoon soup into a blender and blend until smooth. Place in prepared canning jars, leaving a 1-inch head space. Process soup in boiling water bath according to instruction booklet's time table for your altitude.

Creamy Tomato Soup

7 quarts skinned tomatoes
3 medium onions, chopped
3 stalks celery, chopped
12 sprigs fresh parsley
½ cup sugar
3 bay leaves
3 Tbsp. salt
2 tsp. black pepper
14 Tbsp. of flour
14 Tbsp. of melted butter

TO SKIN TOMATOES, PLACE tomatoes in hot water until skins split. Immediately place in cold water. Skins will slip off. Place all ingredients, except flour and butter, in a large stock pot over medium-high heat, stirring frequently. Cook until all vegetables are tender. Carefully spoon soup into a blender and puree until smooth. Return tomato soup puree to stock pot and bring to boil. In a bowl, combine butter and flour. Stir until butter and flour are a smooth paste. Add butter mixture to boiling puree and stir. Place soup in prepared canning jars, leaving a 1-inch head space. Process in boiling water bath according to instruction booklet's time table for your altitude

Dried Beans and Peas

(Hot pack)

4½ cups dried beans or peas
½ tsp. salt or ½ tsp. chicken bouillon granules per
    pint
water to cover beans

STEP 1: WASH AND sort dried beans. Let stand in water overnight. Drain and rinse in morning.

OR

WASH AND SORT DRIED beans. In a large pot, cover beans with water and bring to a boil for 5 minutes. Remove pot from heat and cover. Allow beans or peas to rest for 1 hour. Rinse and drain the beans or peas and leave in pot.

STEP 2: FILL POT with water and boil beans for 15 minutes. Pack into prepared canning jars while beans are boiling hot. Add salt or chicken bouillon granules. Fill jar with bean water, being sure to leave a 1-inch head space. Wipe jar lip clean. Place lid on jar. Place screw band on jar and firmly tighten by hand. Dried beans and peas must be pressure canned according to the instruction booklet's time and pressure table for your altitude.

Fruit Recipes

Canning Syrups

Light Syrup
1 cup sugar
3 cups water

Medium Syrup
1 cup sugar
2 cups water

Heavy Syrup
1 cup sugar
1 cup water

IN A MEDIUM SAUCEPAN, combine sugar with water. Stir over medium heat until sugar is dissolved.

Applesauce

(Hot pack)

10 large apples
2 cups water
1¼ cups sugar
cinnamon to taste

WASH, PEEL, QUARTER, AND core apples. Place apples in a large, thick-bottomed cooking pot. Pour water over apples and cook over medium heat until soft. Place cooked apples in food processor and blend to desired consistency. Return apples to cooking pot and add sugar and cinnamon. Bring to a boil. Pack into prepared canning jars while apple sauce is boiling hot, leaving a 1-inch head space. Wipe jar lip clean. Place lid on jar. Place screw band on jar and firmly tighten by hand. Apple sauce can be boiling water bath processed according to the instruction booklet's time table for your altitude.

Apricots

(May cold pack)

apricots
canning syrup (refer to page 155)

WASH AND HALVE APRICOTS and remove pits. Pack apricots into prepared canning jar. Fill with syrup, leaving a 1-inch head space. Wipe jar lip clean. Place lid on jar. Place screw band on jar and firmly tighten by hand. Apricots can be boiling water bath processed according to the instruction booklet's timetable for your altitude.

Pears

(Hot pack)

pears
canning syrup (refer to page 155)

WASH, PEEL, CORE, AND slice or halve pears. Place pears in a large cooking pot with syrup. On medium heat, boil for 3 to 5 minutes. Pack hot pears into prepared canning jars. Fill with syrup leaving 1-inch head space. Wipe jar lip clean. Place lid on jar. Place screw band on jar and firmly tighten by hand. Pears can be boiling water bath processed according to the instruction booklet's time table for your altitude.

Peaches

(May cold pack)

peaches
canning syrup (refer to page 155)

WASH, PEEL*, PIT, AND slice or halve peaches. Pack peaches into prepared canning jar. Fill with syrup, leaving a 1-inch head space. Wipe jar lip clean. Place lid on jar. Place screw band on jar and firmly tighten by hand. Peaches can be boiling water bath processed according to the instruction booklet's timetable for your altitude.

*PEACH PEELS WILL SLIP if off you dip them for 30 seconds in boiling water and then plunge them into cold water.

Plums

(May cold pack)

plums
canning syrup (refer to page 155)

WASH PLUMS. PRICK PLUM skins, to prevent splitting during processing. Halve plums and remove pit. Pack plum halves into prepared canning jar. Fill with syrup, leaving 1-inch head space. Wipe jar lip clean. Place lid on jar. Place screw band on jar and firmly tighten by hand. Plums can be boiling water bath processed according to the instruction booklet's timetable for your altitude.

Rhubarb

(Hot pack)

1 quart rhubarb
1 cup sugar

PREHEAT OVEN TO 350°F. Wash and cut rhubarb into 1-inch pieces. Place rhubarb into a baking dish and add sugar. Cover dish and bake rhubarb until tender. Pack hot rhubarb and syrup in baking dish into prepared canning jars. Leave a 1-inch head space. Wipe jar lip clean. Place lid on jar. Place screw band on jar and firmly tighten by hand. Rhubarb can be boiling water bath processed according to the instruction booklet's timetable for your altitude.

Cherries

(Cold pack)

cherries
canning syrup (refer to page 155)

WASH CHERRIES AND REMOVE pits or leave cherries whole. Fill prepared canning jars. Fill with syrup, leaving 1-inch head space. Cherries can be boiling water bath processed according to the instruction booklet's timetable for your altitude

Meat, Poultry & Fish Recipes

Roast Beef

(Raw pack or lightly brown in oil with seasonings)

roast beef
seasoning salt

CUT ROAST INTO CHUNKS. Sprinkle with seasoning salt. Pack roast beef in prepared jars, leaving a 1-inch head space. Wipe jar lip clean. Place lid on jar. Place screw band on jar and firmly tighten by hand. Roast beef must be pressure canned according to the instruction booklet's time and pressure table for your altitude.

Hamburger

(Hot pack)

hamburger

IN A MEDIUM FRYING pan over medium heat, brown hamburger. Pack prepared pint jar with browned hamburger, leaving a 1-inch head space. Wipe jar lip clean. Place lid on jar. Place screw band on jar and firmly tighten by hand. Hamburger must be pressure canned according to the instruction booklet's time and pressure table for your altitude.

Ground or Link Sausage

(Hot pack once lightly browned)

sausage

IN A MEDIUM FRYING pan over medium heat, lightly brown sausage. Pack prepared pint jar with browned sausage, leaving a 1-inch head space. Wipe jar lip clean. Place lid on jar. Place screw band on jar and firmly tighten by hand. Sausage must be pressure canned according to the instruction booklet's time and pressure table for your altitude.

Chicken, Turkey, Goose, or Duck breasts

chicken, turkey, goose, or duck breasts*
Salt

Wash breasts in cool water. Cut breasts into pieces and sprinkle with salt. Place breasts in prepared pint jars, leaving a 1-inch head space. Wipe jar lip clean. Place lid on jar. Place screw band on jar and firmly tighten by hand. Poultry must be pressure canned according to the instruction booklet's time and pressure table for your altitude.

*Legs and thighs can also be canned, though times are increased due to the bones.

Salmon, Bass, or Trout

(Raw pack)

salmon, bass, or trout
salt

Wash fish. Fillet fish and remove skin. Pack fish into prepared pint canning jar, leaving a 1-inch head space. Wipe jar lip clean. Place lid on jar. Place screw band on jar and firmly tighten by hand. Fish must be pressure canned according to the instruction booklet's time and pressure table for your altitude.

Pickles, Beets, Relish, and Dressings

WORDS OF WISDOM FROM GRANDMA: Pickles, beets, and relish bottled in the summer should not be eaten until Thanksgiving. Why? You need time for the pickling spices to marinate the vegetables.

Gram's Sweet Pickles

7 quarts sliced cucumbers
1 quart pickling onions
1 quart vinegar
1 Tbsp. mustard seed
3 Tbsp. salt
½ cup sugar

PLACE ALL INGREDIENTS in a large pot and simmer over medium heat for 10 minutes. Drain off liquid. Fill prepared canning jars.

Pickle Syrup

3½ cups vinegar
5¾ cups sugar
2¼ tsp. celery seed
1 Tbsp. allspice

PLACE SYRUP INGREDIENTS IN a large pot and bring to a boil over medium-high heat. Fill canning jars, leaving ½-inch head space. Process according to water bath instruction booklet's timetable for your altitude

Uncle Wanless's Sweet Dills

PLACE CLEAN CUCUMBERS IN ice water for 4 hours. Drain cucumbers. Place in prepared canning jars. Prepare syrup:

PICKLE SYRUP
1 cup water
2 cups vinegar
1 cup sugar
2 Tbsp. salt
6 Tbsp. dill seed

PLACE INGREDIENTS IN A large pot. On medium-high heat, bring to boiling. Fill canning jars, leaving ½-inch head space. Process pickles according to water bath instruction booklet's timetable for your altitude.

Beets

(Must be hot packed)

Beets

Syrup:

  2 cups sugar
  2 cups beet juice or water
  2 cups vinegar
  2 Tbsp. pickling spice
  2 Tbsp. alum

SELECT SMALL, UNIFORM BEETS. Wash carefully, leaving root and 3 inches of tops. Removing roots and tops will cause beets to lose color. Boil beets for 15 minutes or until tender.

TO REMOVE BEET SKIN, place boiling hot beets in cold water. The skin will just slip off with gentle rubbing in your hands. In a large saucepan over medium heat, combine all ingredients, except alum. Simmer for 15 minutes. Stir in alum. Pack prepared jars with hot beets and syrup mixture, leaving a ½-inch head space.

BEETS MUST BE PRESSURE canned according to the instruction booklet's time and pressure table for your altitude.

Bread and Butter Relish

STEP 1:

16 large cucumbers or zucchini
8 large onions
8 green peppers
2 quarts water

GRIND THE ABOVE INGREDIENTS in a food processor or blender. Place in large pot. Add water and boil for 15 minutes. Drain through a sieve.

STEP 2:

1 quart vinegar
5 cups sugar
4 tsp. mustard seed
3 tsp. turmeric
4 tsp. celery seed
2 heaping Tbsp. salt

IN A LARGE POT, combine prepared ingredients from step 1 and ingredients from step 2. Boil for 30 minutes over medium heat. Fill prepared canning jars, leaving a ½-inch head space. Process according to water bath instruction booklet's timetable for your altitude.

Poppy Seed Dressing

1 cup granulated white sugar
2/3 cup rice vinegar
1/4 cup chopped red onion
3 tsp. dry mustard
1 tsp. salt
2 cups safflower oil or olive oil
2 Tbsp. poppy seeds

COMBINE FIRST FIVE INGREDIENTS in a blender. Process ingredients on medium speed until well blended. With blender on a low speed, drizzle oil into mixture. Dressing will become thick and creamy. Once oil is incorporated, stir in poppy seeds. Refrigerate any unused portion in a sealed container.

Taco Salad Dressing

1 cup ranch dressing
1/4–1/2 cup (depending on desired heat) mild salsa
    verde
1/2 bunch chopped cilantro
1/2 lemon, juiced
1 Tbsp. sugar

COMBINE ALL INGREDIENTS IN a blender, processing on medium speed until blended. Refrigerate any unused portion in a sealed container.

Gram's Russian Salad Dressing

1 cup sugar
½ cup vinegar
¾ cup ketchup
¾ cup oil
½ cup grated onion
2 Tbsp. lemon juice
1 tsp. salt
1 tsp. pepper
¼ tsp. garlic powder
½ tsp. dry mustard
1 tsp. horseradish

BLEND ALL OF THE above ingredients in a blender or food processor until smooth. Serve chilled. Refrigerate any unused portion in a sealed container.

Mom's Thousand Island Salad Dressing

1 cup mayonnaise
¼ cup ketchup
2 Tbsp. vinegar
¼ cup granulated white sugar
2 Tbsp. sweet pickle relish
½ tsp. onion powder
Dash of Worcestershire sauce
Salt and pepper to taste

PLACE ALL INGREDIENTS IN a small bowl and whisk until smooth. Refrigerate any unused portion in a sealed container.

Ranch Salad Dressing Mix

¼ cup powdered buttermilk
3 Tbsp. minced dried onions
3 Tbsp. minced dried parsley
1 Tbsp. dried minced chives
½ tsp. salt
1 tsp. sugar
½ tsp. garlic powder
¼ tsp. paprika
⅛ tsp. celery salt
½ tsp. ground black pepper
1 cup mayonnaise
¾ cup milk

PLACE DRY MIX IN an airtight container. When ready to use, blend ¼ cup mix with 1 cup mayonnaise and ¾ cup milk. Refrigerate any unused portion in a sealed container.

Zesty Vinaigrette

1 cup oil
½ cup apple cider vinegar
2 Tbsp. granulated white sugar
3 Tbsp. corn syrup
2 Tbsp. dried bell peppers
1 Tbsp. Italian seasoning
2 tsp. garlic powder
1 tsp. onion powder

MIX ALL INGREDIENTS IN blender and process until smooth. Stir before serving. Refrigerate any unused portion in a sealed container.

French Salad Dressing

⅔ cup ketchup
⅓ cup sugar
½ cup vegetable oil
¼ cup apple cider vinegar
1 Tbsp. Worcestershire sauce
1 tsp. onion powder
¼ tsp. black pepper
¼ tsp. salt

BLEND ALL INGREDIENTS IN a blender or food processor until smooth. Refrigerate any unused portion in a sealed container.

Sweet and Spicy Mustard

1 cup dry mustard
1 cup vinegar
2 raw eggs
½ cup honey
½ –1 tsp. (depending on desired spiciness) prepared horseradish

COMBINE VINEGAR AND MUSTARD in a medium saucepan. Whisk in eggs, honey, and prepared horseradish. Over medium heat, constantly stirring, cook mixture for 8–10 minutes, or until desired consistency is reached. Refrigerate in a sealed container for 24 hours and serve.

Trent's Mayonnaise

1 raw egg*
¼ tsp. salt
1 Tbsp. white vinegar
¾ to 1 cup vegetable oil

PLACE EGG AND SALT in blender. Blend on high for 10 seconds. While blender is on, drizzle oil into egg and salt mixture. When the mixture turns thick and creamy, stop adding oil. Usually, about ⅓ of the oil will be used. Next drizzle in vinegar.

ONCE VINEGAR IS ADDED, add remaining oil until oil no can no longer be incorporated. Refrigerate in a sealed container.

*RAW EGGS HAVE BEEN found to contain Salmonella bacteria. The Centers for Disease Control and Prevention estimates that 1 in 10,000 eggs in America contain Salmonella. We therefore advise any one who would like to eliminate the risk of Salmonella infection to use pasteurized eggs. Pasteurized eggs are commonly found in the refrigerated section of your local grocery store. People in high risk groups such as the elderly, children, pregnant women, and those with a weakened immune system should always use pasteurized eggs when making recipes that call for raw eggs, or if you are planning on eating raw cookie dough.

Trent's Tartar Sauce

1 batch Trent's mayonnaise, substituting lemon
   juice for vinegar
2 Tbsp. minced pickles
2 Tbsp. minced fresh onion
1 Tbsp. minced fresh parsley
1 tsp. dried tarragon
¼ tsp. dried dill
¼ tsp. sugar
⅛ tsp. cayenne pepper

MAKE ONE BATCH OF Trent's mayonnaise using lemon juice instead of vinegar. Stir in remaining ingredients. Chill and serve. Refrigerate in a covered container.

Trent's Sprouting Methods

Seeds

In order to have success with sprouting, you need good quality, organically grown seeds. Do not store your seeds in an oxygen-free environment, or they will not sprout. If you store seeds in sealed containers, do not use oxygen absorbers. If you want the greatest yield possible from your seeds, you will need to air out your seeds every six months. To do this, Trent pours our seeds gently from one bucket to another 2 to 3 times.

We also do not recommend using farm seed, because it may have been pretreated with chemicals. There are many reputable sprouting seed companies to choose from. We encourage people, wherever they live, to support local businesses by purchasing local products. We purchase our seeds directly from www.lifesprouts. com, which is located in northern Utah. We have had good success with their products. We are not promoting Life Sprouts products; we are simply giving you an example of where you can get premium sprouting seeds.

We also suggest that you begin your sprouting adventure by purchasing small quantities of sprouting mixes to determine which seed combinations you and your family will enjoy eating. There are two products from Life Sprouts that we particularly like. The first is a leafy mix called Alfa-Plus-Mix, which contains a combination of alfalfa, clover, cabbage, and radish. We add millet to the leafy mix for nutritional reasons. On occasion, we add more radish seed to the mix to give our salad or sandwich an extra burst of flavor. The second mix that we enjoy is a bean mix called Pro-Vita-Mix. It contains adzuki beans, peas, lentils, mung beans, triticale, wheat, and fenugreek.

You will find that micro-seeds, such as chia seeds, develop a gel-coat during germination that plugs the screen of your sprouting jar. Do not dismay—you can still sprout micro-seeds by placing them on the top of a sprouting bag.

Equipment

We use plastic lids with stainless-steel screens that are designed to fit wide-mouth canning jars. We like them because they come apart, making cleaning easy. Remember, you must tilt the sprouting jars at a 45-degree angle during the sprouting and draining process to allow for optimal aeration and drainage. If you do not have proper aeration and drainage, your sprouts will mildew and spoil.

Sprout draining systems come in all different sizes, shapes, and prices. They can be as simple as a dish drying rack or be an expensive, fancy system sold for a hundred dollars or more. Trent tried a dish drying rack and the sprouting jars would tip and fall. We refused to buy an expensive system, so Trent created his own, using readily available ABS pipe. We have provided you with the materials list, easy-to-follow instructions, and step-by-step pictures in this section of the book, predictably titled, "Trent's Sprout Drainer."

Directions

To start seeds, place them in the bottom of a clean, empty canning jar. Fill jar about ⅔ full of water and let stand overnight. Rinse and drain seeds four times the next morning and place in drainer. For the leafy mixture, we place enough seeds to just cover the bottom of the jar, about 3 tablespoons. The seeds will expand about 20 times their original size as they sprout. For the bean mixture, Trent fills the jar about ¼ full. The beans will expand about four times their original size.

Place the sprouting jars in plain sight where you will not forget to care for them. Sprouts need to be rinsed and drained with fresh water at least every twelve hours. This removes the toxins that are produced as the seeds grow, and it also prevents mildew. Trent rinses our sprouts before leaving for work and after coming home. Rinse your sprouts with fresh water between 65°–70°F—a little cooler than lukewarm. We rinse and drain each jar twice before placing it back in the drainer. Because house temperatures change with the seasons, in the summer, Trent rinses our sprouts with water about 65°F; and in winter the water is closer to 70°F. Sprouts like a room temperature between 65°–74°F. If your house is 80° in the summer, you will more than likely have problems with mildew. Some indirect light is good for sprouts; however, avoid direct sunlight; otherwise, the seeds will cook, mildew, or dry out.

Sprout the leafy mix until tiny green leaves form, about six days. Sprout the bean mix until all the seeds have a tiny white root no longer than the seed itself. This takes about four days. Sprouts produce seed hulls. Hulls are the empty outer seed shell that remain after the seed opens. Don't worry about getting rid of all of the hulls—they just add crunch, like popcorn hulls do after popping. When draining the jars, many of the hulls will stick to the screen. We simply remove the lid after each draining and rinse away the hulls. You can also use a spoon to remove unsprouted seeds and hulls from inside the jar. We don't get too worried about eating some of the hulls. It's extra fiber.

Rotation

If you eat a jar of mature sprouts in three days, remember to start a new one every three days. This will ensure that you will have a continual supply of fresh and nutritious sprouts for your family. Once sprouts are mature, store them in the refrigerator. If you rinse your refrigerated sprouts daily, they will easily keep fresh for a week.

Eating

Rinse the sprouts just before you serve them; they will taste fresher. In the summer, we like to add the bean and leaf mixtures to a lettuce salad at about a 50/50 mix. If you don't have lettuce, just mix the leaf and bean sprouts together and add a little salad dressing. Sprouts are wonderful in winter, when fresh produce is at a premium price. I am always looking for ways to add vegetables to our diet, and breakfast is no exception. During one of our demo group sessions, Tammy Hulse shared a great way to add fresh vegetables to a summer breakfast. Simply scramble eggs and serve them with a side of cottage cheese that has been topped with cucumbers, raw bell peppers, green onions, and bean and seed sprouts. Believe it or not, it's delicious!

How to Build Trent's Sprout Drainer

Materials

The materials required to build my sprout drainer can be purchased at any building supply store or plumbing store. ABS pipe is the black pipe used for sewer systems. The total cost of materials will be around $24 (as of 2009).

Three (3) 3-inch diameter, ABS, 45 degree "Y" connectors
Two (2) feet of 4" diameter, ABS, Schedule 40 pipe
One (1) small can of ABS glue
One (1) 2-inch long screw

TOOLS
hack-saw
knife
drill and bit

Directions:

1. Cut the 4" diameter pipe in half lengthwise.
2. Take one of the 4" diameter pipe half-sections and cut it across-ways into three 7" long sections of half pipe. You will have 3" left over.

3. Cut the 3" remaining half-pipe section so it is 2" wide instead of 3" wide.

4. Cut the rim-flange off of the base of two of the "Y" connectors. This is done so they will fit into the other connectors.

5. Use the knife and round off any sharp edges on all the pieces you have just cut.

6. Glue the three "Y" connectors together in a straight line and let dry.

7. Glue a half-pipe 7" section onto the back of each of the three "Y" connector's 45 degree opening.

8. Make a base support structure by gluing the half-pipe 2" sections on top of one of the cut off flanges.

9. When the glue dries, position the drainer system at a 90° angle in the base support structure.

10. Drill a small hole through one side of the base-support and the drainer pipe flange.

11. Place the screw through this hole to keep the drainer upright in the base support. The base support will provide the elevation support and tilt needed to drain the water into the sink. The screw will keep the structure from twisting sideways and from falling over.

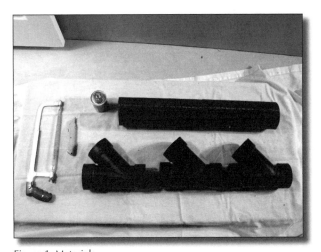

Figure 1: Materials

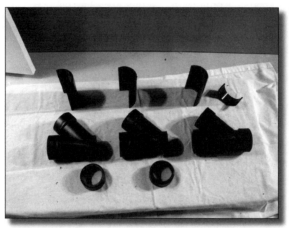

Figure 2: Cuts (see Directions: Steps 1–4)

Figure 3: Base support (see Directions: Step 8)

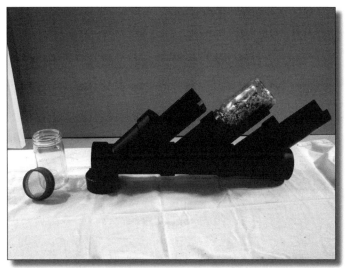

Figure 4: Completed sprout drainer

Chicken Coops and Eggleberries

What's an eggleberry? An eggleberry is what my father calls chicken eggs. Growing up, we had a small flock of about 8 Rhode Island Reds, and in the evening after dinner, my father would ask me if I had gathered the eggleberries. I would giggle, grab an egg carton, and collect the eggs. Now, when my father visits, I ask him what he wants for breakfast, and with a twinkle in his eye, he replies, "Oh, two eggleberries over easy!"

In this section, you will learn what I was taught by my father during our chicken-raising years and how to make your own suburban chicken coop.

How to tell the age of an egg

Using a clear glass bowl of water, submerge the egg. If the egg lies flat on the bottom, it is fresh. If the egg bobs slightly off the bottom of the bowl, it is about a week old. As the egg ages, the larger end will begin to rise. When the egg is about 3–4 weeks old, the egg will stand on its small end with the large end up. If the small end of the egg floats off of the bottom of the bowl, it has spoiled and should not be used.

How to read an egg carton

Egg cartons typically have a "use by" date followed by some numbers. Those numbers are actually a dating system called Julian dates. Typically, the Julian date corresponds to the day that the eggs were packaged. The Julian system begins with number 1 on the first day of January and ends with number 365 on December 31. Eggs typically can last in an egg carton in the refrigerator for about 4–5 weeks without significant loss of quality.

Fresh eggs can last six months!

Yes, you read that right—six months! I have used two different methods for extending the life of my fresh eggs. The first is to freeze them. It's really a simple process.

1. Scramble the raw egg and pour into ice cube trays. Hint: Measure the size of your ice cube trays so you know what size egg will fit in a cube. I have one ice cube tray that holds one large egg in a cube and another ice cube tray that only hold a small egg in each cube. If you forget to test the size of your cubes, 3 tablespoons of thawed whole egg equals approximately one large egg.
2. Once the eggs are frozen solid, pop them out of the ice cube trays and into large freezer bags.
3. Remove air from bag. Exposure to air causes frozen eggs to become tough and rubbery.

When you are ready to use your eggs, place them in a covered bowl in the refrigerator to defrost. I usually do this the night before I need them so that they are ready for breakfast. When eggs are defrosted, they look horrible. Just rescramble them with a fork and use as you normally would. I have found that my frozen eggs taste great for about six months. After six months, the eggs develop a stronger flavor and can only be used in baking.

The second method I use was taught to me by an old shepherd. He would cover his eggs with a thin coating of vegetable shortening. He told me that the vegetable shortening seals off the thousands of pores that are in an egg shell, thus preventing spoilage. I have experimented with this method, and my eggs lasted 5 months in an egg carton in the refrigerator. If you try this method, I would strongly suggest that you test your eggs for spoilage before you eat them or add them to any recipes.

How to cut up a whole chicken

1. Rinse and dry chicken. Place on cutting board breast up with the legs facing you.
2. Pull the leg away from the chicken breast and slice through the skin. You will see where the thigh is attached to the breast.
3. Place the leg flat on the cutting board and slice, in one swift motion, through the meat and the joint.
4. Repeat on other side. Use the same technique for the wings.
5. Firmly hold the chicken breast and slice from the pointy end of the breast where it attaches to the cavity. This will separate the breast from the back of the chicken.
6. Turn the chicken breast over and cut down the middle of the breast. You will now have two halves of breast.
7. You may further cut each half of breast into smaller portions if desired.

The story behind Trent's suburban chicken coop

Fellas, if you happen to marry a woman who grew up with chickens, you may find she has some unusual requests. Here's an example of what I am talking about. Trent and I had been married for eight years when one day it hit me . . . I missed having chickens! In fact, if I ever wanted to feel complete again, I HAD TO HAVE SOME CHICKENS!

Having identified my need, I approached my husband, and using all of my womanly charms, I asked if I could have four chickens. His eyes bugged out of his head, and he promptly said, "NO! We don't live on a farm. Have you forgotten we live in suburbia?"

I responded by pouting out my lower lip, stomping my right foot, and, while placing my hands squarely on my hips said, "Trent, most women ask their husbands for breast implants, jewelry, and fancy vacations! All I am asking for is four chickens." Needless to say, that very night, Trent was on the computer designing the perfect suburban chicken coop for his bride. The program Trent used for designing the chicken coop was Solid Works 2007, SP3.1, Dassault systémes, Concord, Massachusetts, USA.

Suburban Chicken Coop Design

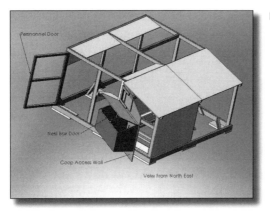

Figure 1: Northeast view

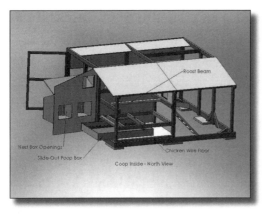

Figure 2: Interior view

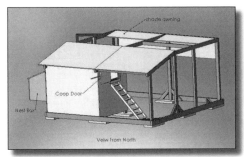

Figure 3: View from North

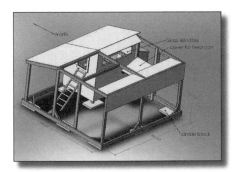

Figure 4: Southwest view

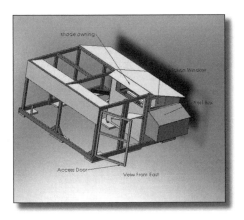

Figure 5: East view, top

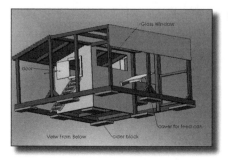

Figure 6: Southwest view, bottom

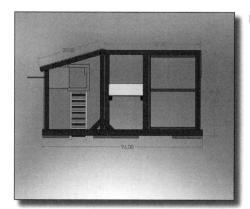

Figure 7: West view, side

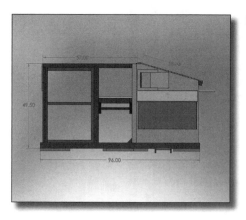

Figure 8: East view, side

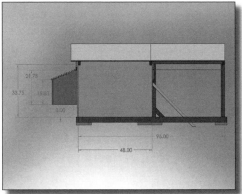

Figure 9: North view, side

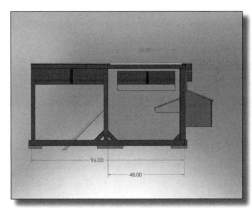

Figure 10: South view, side

Suburban Chicken Coop Details

Exterior Details

The chicken coop system Trent designed has an 8' by 8' square base made of 2" x 4" studs placed on top of flat cinder blocks and raised to about 4' high using 2" x 3" studs. The sides and top of the coop are covered with "hardwire cloth" (rabbit pen wire). The hardwire cloth is attached to the coop studs with staples. The hardwire cloth extends past the foundation and is buried in the ground 12 inches deep. Trent chose a depth of 12 inches to deter burrowing animals from gaining access into the coop. He chose hardwire cloth because the size and the gauge of the squares prevents mice and birds from getting into the coop and raccoons and skunks from chewing through the wire. The square footage provided here is ideal for four large hens. Four hens will lay on the average three eggs a day, unless they are molting, which causes a decrease in egg production.

The pen can be accessed through the door shown in Figure 3. Note the special covering made for the feed can, shown in Figure 4, which keeps the feed dry. Though you can not see this in the drawing, Trent placed a hook under the covering. The hook suspends the feed can so that the food is not kicked over or filled with debris.

The northern one-third of the pen has a roof to keep the hens dry during inclement weather. This opening has a sliding door as well and can be seen in Figure 3. Trent cut a window on the east side of the coop for cross-ventilation. A sliding door with a dowel attached allows me to lock the hens in or out of the coop while cleaning and so on. A wooden ladder allows the hens access to the roost or the laying nests.

The south side of the coop has a large glass window in front of the roosting beam (as seen in Figure 4). Trent has a light bulb and timer wired on the inner south wall of the coop. Hens need at least fourteen hours of light per day to maximize egg production. The large, south-facing glass window allows maximum natural and artificial light to enter the roost. In Utah, there are only three months out of each year when there fourteen 14 hours of natural daylight; therefore, we need supplemental lighting to ensure optimal egg production. The electrical outlet for the 40-watt light bulb and timer also supplies power to a heated watering can in the winter to keep ice from forming.

In order to provide the roosting hens with full sun in the winter and full shade in the summer, Trent angled an awning above the window. During the summer, Trent also covered the south-facing side with a tarp for additional shade.

Interior Details

The nesting boxes are accessed from outside of the pen to make egg gathering easy. The common roof of the nesting boxes is hinged (see Figure 1). Under the roof, the nesting area is divided into two nesting boxes (See Figure 2). Our four hens prefer to share one box, but on occasion, they will switch from one box to the other, so I guess hens like variety too.

The inside of the coop can be seen in Figure 2. The roost beam is by the door and windows. The false floor below the roost beam is covered with chicken wire so droppings will fall into the slide-out tray on the bottom. We put straw in the tray and clean it out as necessary. The holes to the nesting boxes are at the same level as the false chicken wire floor. The entire east side wall is a hinged door that will open to allow the dropping tray to be easily removed. The front of the tray has a handle for convenience.

Trent placed a large auto-feed watering can up on a cinder block in the pen area. If the watering can is not elevated off the ground, the hens will kick debris into the water. Hens require oyster-shell for calcium, so we purchased an oyster-shell dispenser from a local feed store. We would also suggest that you provide your hens with grit, an assortment of fresh fruit and vegetables in the summer and corn in the winter to ensure optimal nutrition and health. We feed our hens pellets from the feed store which contain 28 percent protein and are designed for laying hens. Be sure to provide your hens with dirt so that they may dust for mites.

SECTION THREE:
THE NITTY GRITTY

Weights and Measures

Smidgeon = 1/32 of a tsp.

Pinch = 1/16 of a tsp.

Dash (liquid) = 6 drops

Dash (dry) = $1/8$ tsp.

1½ tsp. = ½ Tbsp.

3 tsp. = 1 Tbsp.

2 Tbsp. = 1 ounce

4 Tbsp. = ¼ cup = 2 ounces

$1/3$ cup = 5 Tbsp. + 1 tsp. = $2^2/3$ ounces

½ cup = 8 Tbsp. = 4 ounces

$2/3$ cup = 10 Tbsp. + 2 tsp.

¾ cup = 12 Tbsp. = 5 1/3 ounces

1 cup = 16 Tbsp. = 8 ounces

2 cups = 1 pint = 16 ounces

4 cups = 2 pints = 1 quart = 32 ounces

4 quarts = 16 cups = 1 gallon

Common Ingredient Substitutions

Allspice, 1 tsp.	½ tsp. ground cinnamon and ½ tsp. ground cloves
Arrowroot starch, 1½ tsp.	1 Tbsp. all-purpose flour OR 1½ tsp. cornstarch
Baking powder, 1 tsp.	⅓ tsp. baking powder and ½ tsp. cream of tartar OR ¼ tsp. baking soda and ½ cup sour milk or butter-milk (decrease liquid in recipe by ½ cup) OR 1 tsp. baking soda and ½ Tbsp. vinegar or lemon juice used with milk to make ½ cup (decrease liquid in recipe by ½ cup)
Beef broth or chicken broth, 1 cup	1 bouillon cube dissolved in 1 cup water
Butter, 1 stick	7 Tbsp. vegetable shortening
Buttermilk, 1 cup	1 cup milk minus 1 Tbsp. and 1 Tbsp. lemon juice or vinegar (allow to stand 5 to 10 minutes)
Ketchup, 1 cup	1 cup tomato sauce, ½ cup granulated white sugar, and 2 Tbsp. vinegar
Chili sauce, 1 cup	1 cup tomato sauce, ¼ cup brown sugar, 2 Tbsp. vinegar, ¼ tsp. ground cinnamon, dash of ground cloves, and a dash of allspice
Chocolate, semi-sweet, melted 6 ounce package	2 squares unsweetened chocolate, 2 Tbsp. shortening and ½ cup granulated white sugar
Coconut, 1 cup grated	1⅓ cups flaked coconut

Corn syrup, 1 cup	1 cup granulated white sugar and ¼ cup liquid (use whatever liquid is called for in recipe) OR 1 cup honey
Cornstarch, 1 Tbsp.	2 Tbsp. all-purpose flour OR 2 Tbsp. granular tapioca
Cream, whipped, 1 cup	Chill a 13 ounce can of evaporated milk for 12 hours. Add 1 tsp. lemon juice. Whip until stiff.
Heavy cream, used for cooking, 1 cup	⅓ cup butter and ¾ cup milk
Light cream, used for cooking, 1 cup	⅞ cup milk and 3 Tbsp. butter
Egg yolks, 2	1 whole egg
All-purpose flour, 2 Tbsp.	1 Tbsp. of cornstarch OR arrowroot OR potato starch OR quick cooking tapioca
All-purpose flour, 1 cup	1½ cups bread crumbs OR 1 cup rolled oats OR ⅓ cup cornmeal and ⅔ cup flax seed meal
Self rising all-purpose flour, 1 cup	1 cup minus 2 tsp. all-purpose flour and 1½ tsp. baking powder and ½ tsp. salt
Honey, 1 cup	1¼ cups granulated white sugar and ¼ cup liquid
Lemon juice, 1 tsp.	½ tsp. white vinegar
Lemon zest, 1 tsp.	½ tsp. lemon extract
Mayonnaise, 1 cup	½ cup yogurt and ½ cup mayonnaise OR 1 cup sour cream OR 1 cup cottage cheese pureed in a blender

Milk, 1 cup	1 cup yogurt
Mushrooms, 1 pound fresh	1 (12-oz.) can mushrooms, drained OR 3 ounces dried mushrooms
Sweetened condensed milk, 1 cup	⅓ cup evaporated milk, ¾ cup granulated white sugar, and 2 Tbsp. butter. Heat until butter and sugar are dissolved. OR 1 cup plus 2 Tbsp. powdered milk to ½ cup warm water. Mix well. Add ¾ cup granulated white sugar and stir until smooth.
Whole milk, 1 cup	1 cup reconstituted nonfat dry milk and 2 tsp. butter. OR ½ cup evaporated milk and ½ cup water.
Sour cream, 1 cup	1 cup plain yogurt
Granulated white sugar, 1 cup	1 cup corn syrup (decrease liquid in recipe by ¼ cup) OR 1⅓ cup molasses (decrease liquid by ⅓ cup) OR 1 cup brown sugar OR 1 cup honey (decrease liquid by ¼ cup) OR 1¾ cup powdered sugar, packed.
Tomatoes, 2 cups chopped	1 (16-oz.) can
Tomato juice, 1 cup	½ cup tomato sauce and ½ cup water
Tomato sauce, 2 cups	¾ cup tomato paste and 1 cup water
Yeast, 1 Tbsp.	1 package (¼ ounce)
Yogurt, 1 cup	1 cup buttermilk OR 1 cup cottage cheese, pureed OR 1 cup sour cream

Food Equivalencies

BEANS
1 cup dried = 2 to 3 cups cooked beans
2 cups cooked = 1 (16-oz.) can beans, drained
1 pound dried beans = 2 cups dried beans

CHEESE
1/3 cup softened cream cheese = 3 ounce package of cream cheese
1 cup cottage cheese = 8 ounces cottage cheese
2 cups shredded cheese = 1 (8-oz.) solid piece of cheese
1 cup bleu cheese crumbles = 1/3 pound bleu cheese
1 cup grated Parmesan cheese = 1/2 pound piece or 3 ounces of already grated Parmesan cheese

CRUMBS
1 cup soft bread crumbs = 2 slices old bread
1 cup dry bread crumbs = 3/4 cup cracker crumbs
1 cup graham cracker crumbs = 12 graham crackers
1 cup soda crackers = 20 soda crackers
1 cup vanilla wafer crumbs = 22 vanilla wafers
1 cup corn flake crumbs = 2 1/2 cups corn flakes
1 cup chocolate wafer crumbs = 26 chocolate wafers

GARLIC AND HERBS
1 tsp. dried herbs = 1 Tbsp. fresh herbs
1 tsp. powdered garlic = 1 Tbsp. fresh garlic
1 large clove of fresh garlic = 1 tsp. fresh garlic

MEAT, POULTRY, AND FISH EQUIVALENCIES

1 pound of fresh meat will fit in a 1 pint jar

2 pounds of fresh meat will fit in a quart jar

1 pound of fresh fish will fill a pint jar ¾ full

1 can of chicken = 1 pound of fresh chicken = 1 pint of bottled chicken

1 can of hamburger = 1 pound of fresh hamburger = 1 pint of bottled hamburger

3 to 4 pound roast = 1 quart of beef

VEGETABLE EQUIVALENCIES

3 cups dried sauce potatoes = 2 large fresh sauce potatoes

1 Tbsp. dried celery = 1 fresh stalk of celery

2 Tbsp. dried carrot = 1 fresh carrot

2 Tbsp. dried onions = 1 medium fresh onion

¼ cup dried bell peppers = ½ cup fresh bell peppers

Recipe Index

W

White Chicken Chili 43

Y

Yogurt 135
Yogurt Cheese 136
Yummy Yammy Casserole 75

About the Authors

Michelle Snow is finishing up a PhD in public health at the University of Utah. Michelle has published over twenty articles in peer-reviewed professional journals and has presented her public health research at national and international conferences. Michelle's hobbies include gardening, hiking, cooking, crocheting funky blankets, snow skiing, and jet skiing. She is married to Trent Snow.

Trent Snow served an LDS mission in Taiwan. He graduated from Brigham Young University with a Bachelor of Science degree in electrical engineering technology. Trent has been employed as an electrical engineer at Hill Air Force Base for the past twenty-three years. Trent enjoys puttering around in the kitchen with Michelle and inventing kitchen gadgets that are practical, economical, and easy to construct. Trent enjoys snow skiing, traditional bow making, identifying edible plants, fly fishing, cheese making, and being

Michelle's taste tester; in fact, the nick name his family has given him is Sir Taste-A-Lot!

Michelle, Trent, and their five children, Rachel, Tyler, Austen, Bryson, and Adam, happily reside in Utah. They are kept company by their dog, Lady Isabella of Bromfield, four pet chickens, Hog Mama, Gertie, Argee, and Puffy Checks, and two cats, Senorita Peachalina Gato and Archimedes.

You can contact Michelle by going to her website:
www.michelle-snow.com
She would love to hear from you!